Pittsburgh-Konstanz Series in the Philosophy and History of Science

Science, Reason, and Rhetoric

EDITED BY Henry Krips,
J. E. McGuire, and
Trevor Melia

University of Pittsburgh / Universitätsverlag Konstanz

Published in the U.S.A. by the University of Pittsburgh Press, Pittsburgh, Pa. 15260
Published in Germany by Universitätsverlag Konstanz GMBH
Copyright © 1995, University of Pittsburgh Press
All rights reserved
Manufactured in the United States of America
Printed on acid-free paper

Library of Congress Cataloging-in-Publication Data
Science, reason, and rhetoric / edited by Henry Krips, J.E. McGuire,
 and Trevor Melia.
    p.   cm.—(Pittsburgh-Konstanz series in the philosophy and
history of science)
    ISBN 0-8229-3912-6 (alk. paper)
    1. Science—Methodology.   2. Science—Philosophy.   3. Rhetoric—
Philosophy.   4. Rationalism.   I. Krips, Henry.   II. McGuire, J. E.
III. Melia, Trevor.   IV. Series.
Q175.S4225   1995
501—dc20   95-33117
                                                                    CIP

Die Deutsche Bibliothek-CIP-Einheitsaufnahme
Science, reason, and rhetoric / ed. by Henry Krips ...—
Konstanz: Univ.-Verl. Konstanz; Pittsburgh, Pa. : Univ. of
Pittsburgh Press, 1995
   (Pittsburgh-Konstanz series in the philosophy and history of
   science; 4)
   ISBN 3-87940-533-6
NE: Krips, Henry (Hrsg.); GT

A CIP record for this book is available from the British Library.

Contents

Introduction

The vitality of the present essays, together with their commentaries, attests to the burgeoning interest in the rhetoric of science. Studies of this important aspect of scientific activity have increased in number significantly over the last decade. Careful attention to the essays gathered here will indicate that the rhetorical prospective brings insights to the study of science not captured by history, sociology, philosophy, anthropology or heremeneutical analysis; that is, rhetoric can occupy a "clearing" that is obscured in these other approaches to science studies. These approaches have "essentialized" their metadiscourses by turning the object of their study—science—into yet another instance of their hegemonic vision. Thus, sociologists see science as reducible to social parameters, historians to the cultural contexts within which scientific practice comes into being, and textual critics reduce science to the protocols of its representative texts. On the other hand, while rhetorical case studies of science are desirable, and to be encouraged, they face a danger of fitting the tools of analysis to the protocols of the case study itself. If the object of inquiry dictates the method of inquiry, the interpretive resources required to bring the object of inquiry to light can be severely compromised if not set aside.

Consider three current approaches to the rhetoric of science. (1) On the first approach, science is not viewed as a truth-producing enterprise. Even if truth were available we might not be able to recover it or to communicate it. Given that our cognitive resources are limited to the extent that truth may not be easily distinguished from falsity, other more accessible epistemic goals should be projected and maintained. Therefore, science should promote strategies which best facilitate inquiry into the natural world, and develop tropes of communication which maximize contact with audiences best placed to support and advance the scientific enterprise. Rhetoric plays a manifest role in scientific communication, in acts of scientific persuasion, and in the structuring of the various audiences of science. Accordingly, the first approach creates a global clearing for rhetorical analysis by reconstituting science as a particular form of human praxis which stands apart from the thematics in other fields of science studies. Essentially it makes science a phenomenon safe for rhetorical analysis. (2) Equally hegemonic, the second approach also construes science *sub specie rhetoricae*. However, instead of linking the dynamics of science to the sophistical tradition of inquiry, the second approach reduces the study of science either to a species of textual analysis in which rhetorical strategies play a central role, or subsumes science under a sociology translated into the categories of rhetorical analysis. On this alternative approach, rhetorical analysis buys into the protocols of the metadiscourse in which it "grounds" itself and accepts thereby any essentializing tendencies inherent in that discourse together with its unexamined assumptions. (3) The third approach eschews any attempts to transform the discourse of inquiry into a form that is friendly to rhetorical imperatives or to treat science as a particular praxis continuous with the sophistical tradition of rhetoric. It establishes a clearing by indicating the multifarious ways in which rhetorical dimensions operate in the exercise of scientific practices. At the outset this approach forbears any premature disclosure of techniques and strategies which may be germane to forging a continuing understanding of the dynamics of the scientific process. It maintains a pluralistic and antihegemonic stance. This does not mean that it grounds itself naively in the "actualities" of science, or that it avoids theorizing the objects of its inquiry. It does mean that an attempt is made to integrate rhetorical elements into a larger whole that contains as well accounts of science that are patently not rhetorical.

The essays presented in this volume fall under the third alternative. They stress the variety and complexity inherent in the production of scientific knowledge and also in the attendant human contexts within which science is made and established. As a whole, the papers range from a concern with rhetorical theory to a concern with cases, including theory–case or case–theory. The essays of S. Toulmin, P. Kitcher, J. E. McGuire and T. Melia, and S. Fuller address issues of theory; those of R. Feldhay, T. J. Pinch, R. P. Newman, and J. A. Campbell present case studies; while those of J. Lyne, H. Krips and G. Beer concern the interface between theory and case. Apart from those of McGuire and Melia, Fuller, Krips, Massey, and Newman, each of these essays was presented at the University of Pittsburgh in a conference entitled "Science, Reason, and Rhetoric" on the occasion of the formal inauguration of the University's new program in the rhetoric of science.

Taken together, the contributors represent a wide range of disciplinary competency, but the individual contributions converge significantly on the central issues of the rhetoric of science: the constitution of scientific authority; the role of discursive devices in the production and establishment of scientific knowledge; the production of scientific discourse especially in settings of contestation and controversy; the interface between the rhetorical and the cognitive and the rational and the probative; and, lastly, the role rhetorical strategies play in the recasting of discourse to fit the receptivity of differently constituted audiences.

Toulmin connects rhetoric and science and communication and philosophy. He shows that rational and rhetorical elements in scientific thought and communication legitimately overlap and share common ground. In Toulmin's view, we should cease to proceed (as we have since the seventeenth century) on the assumption that the rational and the rhetorical are inherently polar opposites. A properly constituted rhetoric of science, he argues, will help in challenging the frontiers between the central categories of our critical reflection on the construction of science and its cultural role in human understanding.

For Toulmin, style and content are inseparably connected in both the production and communication of scientific knowledge. Just as no act of communication is independent of its mode of representation, no language is logically transparent and rhetoric-free. Scientific knowledge, therefore, if appropriately constituted, possesses epistemic goodness not just because it satisfies the formal characteristic of rational argument, but more importantly because it possesses probative power; that is, it persuades and convinces its perspective audiences. In Toulmin's view, the constitution of human thought and communication involves varying proportions of logic, reasoning and rationality. Of their very nature, then, they embrace at once both theorizing and practice as well as reasonableness and persuasion. Here, he suggests, is a significant clearing for rhetorical strategies, a selection of which he analyzes with telling effect. In motivating his position Toulmin provides us with a compelling invocation of Aristotle's *Organon* in which logic, rhetoric, the topics (the codification of patterns of reasoning), and dialectic are developed as interconnected aspects of the probative "reasonableness" of human thought and action.

To find a *via media* between extreme positions of debate often forges a balanced perspective. Such an account is proposed by Kitcher in regard to scientific rhetoric. For him, science is not a "rhetoric-free zone." The linguistic forms employed in science and the rhetorical devices that persuade scientific communities are an integral part of the enterprise of science. Nevertheless, Kitcher resists the rampant over-rhetorizing of science that has been fashionable of late. Tropes of discourse have causal consequences and help to shape the formation of belief and to forge cognitive attitudes. For Kitcher, no essential opposition exists between the rhetorical and the cognitive. Indeed, rhetorical strategies function substantially in science with respect to the business of encouraging valuable epistemic performances and in forming useful beliefs about the world.

Kitcher surveys and analyzes some of the ways in which rhetoric serves to promote the quest for true explanations in science and helps to negotiate our manifest cognitive shortcomings. In this regard he tackles the hard case of mathematical proof. Proofs in mathematics are not just formal contributions embodying logically transparent language; rather, standard mathematical proofs are replete with rhetorical devices which facilitate our cognitive grasp of the mathematics and our ability to see why the proof works. Thus, even in mathematics our cognitive imperfections render it impossible simply to construct rhetoric-free proof structures.

Noting that a rhetoric of science ought to embrace both public discourses of science as well as social dynamics that stand behind them, Kitcher concentrates on the rhetoric of two sorts of published scientific documents: the research article that reports a new experimental finding and the article or book that proposes a new theoretical perspective. With respect to the first type of writing Kitcher discusses the various rhetorical strategies used by researchers to establish their proficiency, expertise and authority within the research community. The business of establishing a new perspective, however, is far more taxing. It involves breaking down deep-seated habits of thought. In particular, new proposals often include sets of inconsistent claims which lead to explanatory loss when one or more of the claims are eliminated by proposed strategies for overcoming inconsistency. Here in contexts of scientific controversy and theoretical change, Kitcher shows how rhetorical strategies contribute to the cognitive tasks at hand and to the formation of innovative authority within the scientific community. To illustrate his analysis he discusses the making of innovative science at the hands of Lavoisier and Darwin.

Arguing that science is a set of situated and constrained practices that employs various sorts of representational devices, McGuire and Melia urge a proportionalizing rhetorical perspective on the interrelated discursive and representational practices of science. They find an important moral for the rhetoric of science in the fact that the theories of science, as well as their methods, norms, and goals, have changed, and are changing, through time. If change in science is in key part characterized by a self-correcting fallibilism, any rhetorical analysis that fails to reflect the proportionalizing strategies of the discourses of science will be wide of its mark. This is not to say that science does not entrench and perfect its constitutive vocabularies. In situations of contestation, however, these vocabularies are significantly disciplined as scientific practice moves to negotiate controversy. After all, as Burke and Geertz point out, a constitutive connection lies between human praxis, science included, and the symbolic structure of human cognitive action. Accordingly, rhetorical, as well as other thematic analyses, ought to reflect these dynamics rather than impose their own hegemonic impulses on the object of their inquiry—scientific practice.

McGuire and Melia examine three of the central hegemonic reductions current in metadiscursive analysis: sociology of scientific knowledge, realism and antirealism, and the ethnographic reduction of science to inscriptional devices. Each of these essentializing accounts of science has sought to impose its perfecting vocabularies on science itself. What is needed to ground a balanced rhetorical analysis is a proportionalized relation between these and other hegemonic positions. McGuire and Melia also note that these modes of representation generate from within an internal recalcitrance which disciplines their pretensions, and encounters, in varying degrees, real world limitations.

What does the "rhetoric" in "rhetoric of science" add that cannot already be gotten from literary criticism, discourse analysis, hermeneutics, or cultural studies, not to mention philosophical relativism? To answer this question, Fuller proposes a Strong Program in the Rhetoric of Science, a conceptual framework and research agenda loosely modeled on D. Bloor's Strong Programme in the Sociology of Scientific Knowledge. Fuller distances this program from philosophical, literary, historical, and sociological studies of science. His ultimate goal is not to propose yet another discipline, however, but to bring to light and criticize certain core theoretical assumptions in these and other approaches to science that have been increasingly suppressed with the rise of the case study method.

More specifically, Fuller argues that the rhetorician of science should forswear any allegiance to such philosophical positions as relativism, realism, and rationalism because they presuppose a conception of language that is unbecoming to the rhetorician. As opposed to the textualist tendency in literary studies (and much of what currently passes as "rhetoric of science"), Fuller urges that the rhetorician of science attend to the reception history of scientific texts, regardless of the sloppy reading habits it may reveal. In addition, Fuller warns the rhetorician of science against the mythic tendencies of narrative history, which may legitimize a certain humanistic or scientific perspective, while it downplays the element of timing, or *kairos,* that is crucial to rhetorical sensibility. Fuller also faults sociologists for fostering forms of knowledge that do not necessarily enable its possessors to transform their social situations. Finally, traditional historians and philosophers of science should be challenged by Fuller's final two questions that place considerations of criticism and audience squarely at the center of any adequate explanation of the modern scientific enterprise.

Lyne argues that just as the "received view" of science (the syntactic theory) has been critiqued in terms of theories inspired by the biological model (semantic theories), both types of theory are destined in turn to come up against limits that are the result of their failure to account for the programmatic dimension of scientific inquiry. From Lyne's rhetorical perspective the "pragmatic" involves "the situated management of meaning where one is trying to make a persuasive case." Such an approach pays much more attention to the mediating effects of language generally and in particular does not reduce the role of metaphor to heuristic only. Rather, metaphor is seen not only as a mechanism whereby the highly technical arguments of science are rendered accessible to popular audiences but also the means whereby the insights of the specialized sciences are conveyed across disciplinary boundaries. Trouble develops, however, when the technical vocabulary of one specialty is transferred to another and the terminology is assumed to retain its evidentiary warrant when in fact the transaction is merely metaphorical and heuristic. Thus, according to Lyne, whenever "gene talk" from genetics is used in sociobiology, "Genes as heuristic becomes genes as warrant. A biological perspective becomes a bio-logical claim." The difficulties engendered by this type of confusion are not resolved by a program simply to identify and to root out the metaphors in science. Such programs, typical of the older syntactic theories, by resorting to a diadic conception of language (intention and extension, denotation and connotation, sense and reference) are led to ignore the embedment of metaphorical language in science and also much of the rhetorical dimension of the enterprise. Lyne avers that C. S. Peirce's pragmatic theory of science in giving a triadic account of signification offers a better chance to explain scientific langauge and communication. Lyne thus applies Peirce's familiar distinction between iconic, indexical and symbolic signs to various aspects of "gene talk" to tease out the rhetorical component in such talk. In so doing Lyne also uses Peirce's approach to shed light on the problem of the continuity of reference of scientific terminologies that has troubled philosophers for some time.

If any piece of scientific discourse could be said to be innocent of ideological and rhetorical dimensions it would be von Neumann's *Mathematische Gründlagen der Quantenmechanik*. By applying insights of Althusser and Lacan, most particularly to a careful reading of the *Gründlagen*, Krips shows otherwise. Mobilizing a familiar continental tactic, Krips concentrates on the "margins" of the text (translator's preface, footnotes, historical introduction) to show tensions between the communal norms of science (the Bacon-Merton version) and the practice of conferring honors and awards on individual scientists for original contributions to science. Krips points out that the Bacon-Merton norms of science are ideological in both the traditional Marxist sense that they misrepresent an underlying social reality and also in the Althusserian sense that they contribute to the "constitution of the subject" who lives this (mis)representation. Seen in this light von Neumann's problem is that the *Gründlagen* must convince the skeptical audience of physicists that also constitutes the very source of communal authority and credibility for his thesis. To accomplish this tour de force von Neumann represents what is actually a critique of his contemporaries generally, and a reproach to Dirac particularly, as a text aimed at educating the otherwise well-informed novice in the mysteries of quantum theory (QT). What is at issue is the romantic-aesthetic account of QT offered by Dirac and largely satisfactory to physicists versus the mathematically rigorous account more consistent with enlightenment ideology, profered by von Neumann and resisted by his contemporaries. By exploiting the distinctions between author and narrator and between narratee and implied reader, and tracing the tensions between author and translator, Krips shows how these ideological conflicts resonate in the text of the *Gründlagen*. The work not only reflects and distorts images of the ideological field in which it was produced but also is itself a site for the constitution of ideology. In support of this perception Krips offers a Lacanian analysis of von Neumann's text wherein the structural similarities between the subject's sense of "self-lack" and the ideological inconsistencies in his cultural environment are productive of ideology via a Lacanian "creation of desire." By scrutinizing von Neumann's work in such terms, Krips not only throws light on the formidable *Gründlagen* but also reduces the opacity of the literary and theoretical ideas he employs.

In a virtuoso reading of Eddington's three popular works on science, *The Nature of the Physical World* (1928), *New Pathways in Science* (1935), and *The Philosophy of Physical Science* (1939) (see references at the end of chap. 11), Beer reminds us just how much the physicist influenced his contemporaries. By the time he led the famous 1919 expedition to Brazil, whose study of the solar eclipse of May 29 was taken to verify Einstein's predictions, Eddington's scientific reputation had already been established. The broad cultural influence that he had achieved by the 1930s was largely the result of the three popular books, all of which were based on lecture series. Some of the lectures must be counted among the earliest broadcasts to a general radio audience on such seemingly exotic and complex topics as the general and special theories of relativity and quantum theory. Beer is more concerned, however, with emphasizing Eddington's influence on such literary notables as Virginia and Leonard Woolf, W. Empson, and W. Benjamin. She points out how the religious sensibilities of this "shy Quaker" rendered Eddington more accepting than his scientific colleagues of the "absence," indeterminacy, and fallibilism that characterized the physical theory of the 1930s. Eddington thus viewed the task of physics as less a matter of decoding a given "book of nature" than of tracing the multiple epistemic implications of the constructed and tautologous narratives of science. Beer shows that the workings of the resulting spirit of play (Derrida's *Jeu*) in Eddington's writing go back to his earliest schoolboy essays. Perhaps this aspect of Eddington's style rendered his work accessible and attractive to the literary notables of the 1930s. Eddington's attack on "substance" thinking obviously lent impetus to Virginia Woolf's aspiration to "insubstantise" the "real"; and that his alternative "Waves! Waves! Waves!" may find resonances in Woolf's novel *The Waves*. However, these are only a few apercus to be gained from a careful reading of Beer's highly nuanced essay that probes the larger questions faced by the rhetoric of science, "[H]ow and to what extent does individual experience go to compose theory? and can scientific thinking give a purchase on communal ideology as well as being informed by it?"

Feldhay explores Galileo's controversy with the Jesuits on the nature of sunspots. In particular she traces the emergence of a new Galilean discourse within a traditional and antagonistic field of scientific discursive practices. She argues that Galileo's discourse concerning sunspots contains no essential separation of the rhetorical from the "scientific," the persuasive from the apodictic, or the analogical from the probative. On the contrary, these elements are part of Galileo's enterprise for establishing his discourse, setting its boundaries, and legitimizing its scope. In particular, Feldhay shows how analogies, metaphors and other rhetorical devices feature in Galileo's attempt to forge a new discourse which captures his vision of "philosophical astronomy." In this exercise, Galileo not only confronts the standard problem of constituting the sunspots as objects of scientific discourse, but negotiates as well his view of the inherent limitations of the discursive practices of traditional "mathematical astronomy," which forces Galileo to reject explicitly the institutional boundaries between the mixed and the mathematical sciences, and between astronmy and natural philosophy. In a more positive vein, it forces him to fix the boundaries of his astronomical discourse while at the same time establishing the authority of the individual to read the language of nature in the face of the authority of institutional consensus and tradition. As Feldhay interestingly shows, this impels Galileo to use important rhetorical maneuvers in the act of constituting the discursive framework of his science.

While humor may on occasions have a decisive rhetorical effect in persuading voters to support a politician, customers to buy goods, and so on, Pinch argues that humor may play a similar role in the scientific marketplace for new ideas. In arguing for this conclusion, he draws upon his earlier work coauthored with C. Clark which identified the rhetorical forms employed by English back-street market traders. For example, as part of their pitch, the traders skillfully developed contrasts between the "high worth" and "low price" of their goods, " 'ere ah don't want four pound, two pound, one fifty, an' not even one twenty . . ." Pinch then draws an explicit parallel between the market traders' conversational ploys and those of scientific conference presentations. His first examples from the sciences are drawn from a speech by the ethnomethodologist Garfinkel and a presentation by a health–care economist. Both speakers use a rhetorical device of the forms "it's not an *A,* it's not a *B,* it's a *C,*" which has the same form as some of the lines used by the market traders.

Pinch draws his second set of examples from the cold fusion debate. He argues that humor played a key role in two public presentations which were instrumental in the short history of this controversy— Pons's address of 12 April 1989 to the American Chemical Society meeting in Dallas (the "Woodstock of Cold Fusion"), and three weeks later the presentations by Koonin and Lewis to the American Physical Society meeting at Baltimore. Pons scored a palpable hit with his audience by wittily contrasting the vast but mostly unsuccessful efforts by physicists in the area of cold fusion with the fairly modest efforts of chemists which (he argued) had already enjoyed some measure of success. In the second meeting, however, Koonin and Lewis turned the tables on the chemists by drawing a humorously oblique contrast between the "good science" of physicists and the sloppy, even fraudulent efforts of the chemists Pons and Fleischmann. In these contexts, humor served two main rhetorical effects: (a) It created solidarity among members of the audience and with the speaker; and (b) it allowed the speaker to make serious allegations without the need to defend them ("cheap shots").

Pinch's essay not only provides compelling instances of the use of rhetoric in public argument among scientists, but also points to significant ways in which the notion of scientific rhetoric must be broadened to accommodate various "lower" forms of persuasion.

Newman explores the circumstances attendant upon the reception of A. Wegener's celebrated drift theory in the U.S. He points out that many historians of the Wegener debate trace the early American hostility to the thesis as a more or less rational outcome of the symposium on the topic held in 1926 by the American Association of Petroleum Geologists. In a detailed examination of the correspondence and documents in the case, Newman shows that far more influential was the publication in 1928 of what purported to be the paper given at the New York conference. In fact, since the original invitation to the conference drew only one mobilist contributor, van der Gracht of Holland, who found himself confronted by the nearly uniform permanentist sentiment that characterized the geological elite in the U.S., the published volume includes a number of papers from scholars who did not speak at the 1926 conference. Of the six American manuscripts submitted to the expanded volume, five supported the antimobilist sentiments previously expressed by the American conference speakers (all of whom were former or future presidents of the Geological Society of America). The unanimity of the American response to the Wegener thesis has been attributed by some to a fear of resurgent catastrophism or to genuine concern about evidence brought by competing theories. Newman sees a much more prosaic reason, namely, the influence of the Mandarins of American Geology led by T. C. Chamberlin and later by his son Rollin, both chairs of the University of Chicago's Department of Geology and editors of the *Journal of Geology* and both in the grip of a powerful orthodoxy. The irony is that T. C. Chamberlin's most enduring single publication, available as late as 1977 as a reprint from the American Association for the Advancement of Science, was a stern warning against the dangers of the tyranny of dominant theories.

The central point that Campbell argues is that the *Origin* is arranged, partially at least, because Darwin is engaged in a rhetorical reconstruction of the *vera causa* principle to meet the rhetorical needs of persuasiveness in his narrative presentation. Darwin is actively using tradition to overcome tradition by rhetorically reconstituting his central metaphor, "natural selection," and his tactical stocking-house, the "*vera causa*" principle. This strategy allows Darwin to reconstitute rhetorically the traditional discourses of natural theology to fit adaptively the contours of his own innovative theory of selection.

Campbell considers the rhetorical and metaphorical structure of Darwin's 1837 entries in his notebooks (lettered "B" through "E"). His focus is Darwin's use of Malthusian arguments and the rhetorical strategies employed by Darwin to transform Malthus's argument for population stasis into one of evolutionary change. After dealing with several of Darwin's attempts to place rhetorically the trope of "natural selection," Campbell turns to the *vera causa* principle. He argues that

Darwin departs from linear logic—the independent existence of the explanatory cause, its competence to account for the change to be explained, and the responsibility of the cause for affecting organic change—in the interest of maximizing the internal protocols of his narrative selection. Thus the "prudential logic" carried by Darwin's rhetorical strategies is essential for portraying the cognitive potential of his theory and for adapting it to target audiences rooted in the culture of Victorian science and society.

Each chapter addresses science as a form of human *praxis*. That is, science is seen as a special type of human activity, a self-correcting activity, which carries with it its own ever-changing norms, aims, and goals. On this conception, science is a situated set of discursive and representative practices. It is, in short, a form of human symbolic action, a term coined by Burke and appropriated and developed by writers such as Geertz and P. Ricoeur. For these writers, human beings are symbol made and symbol making. Symbolicity, however, is not reduced to linguistic structures alone, but refers more generally to the human condition of dwelling in a universe of cultural meanings constitutive of its modes of self-presentation.

So construed, science is a social form possessing a recognizable symbolic structure. This structure, inextricably connected with the praxis of science, reflects, expresses, and represents the inner dynamics of that praxis itself. In other words, scientific praxis possesses always and already a symbolic dimension. This means that scientific practice, as manifested in symbolic action, already possesses rhetorical dimensions and invites rhetorical analysis. Most importantly, it invites modes of analysis that are sensitive not only to the ways in which rhetorical strategies and devices function in the discourse of science, but also to the ways in which larger cultural formations may or may not impinge on the practice of science. At a more particular level, this orientation calls attention to those sites of contestation which are often cradles of scientific innovation and productive of new patterns of scientific knowledge. However, it does so in a way that does not decontextualize the production of knowledge, and that helps reveal the intrinsic connection between representation and intervention and between the discursive and the nondiscursive in scientific practice.

We offer this volume with the aim of showing the importance of rhetorical dimensions in the enterprise of science. However, we also insist that the complexities of science are not reducible to either rhetorical theory or to rhetorical practice. Let rhetorical analysis join with other modes of analysis to illuminate one of the great achievements of the human spirit. It is here in the endeavor of this joint enterprise that a significantly new rhetorical understanding will emerge.

Henry Krips
J. E. McGuire
Trevor Melia

Science, Reason, and Rhetoric

1

Science and the Many Faces of Rhetoric

Stephen Toulmin
The Center for Multiethnic and Transnational Studies, University of Southern California

For much of the twentieth century, philosophers of science concentrated largely on formal aspects of the logic of induction, such as confirmation theory, so that any proposal to extend the discussion into the rhetoric of science is a breath of fresh air. As little as ten years ago, relations between American departments of philosophy and communication were still antagonistic. For example, the closing of the Dartmouth speech communication department was welcome news to the philosophy department. So it is worthwhile for us to begin by asking how far, and in what ways, questions about the rhetoric of science overlap traditional questions in philosophy of science. For our debate to be fruitful, let us clarify the reasons why, in the heyday of logical empiricism, "rhetoric" was not just neglected, but scorned.

An initial point should be made clear: We could raise questions in the rhetoric of science without challenging in any way the bases of earlier philosophy of science. We might, for example, continue to take for granted that the *rational* task of the natural sciences is to refine, improve, or replace the concepts and beliefs current in any field, and—*after that rational task has been carried through*—inquire how the authors of such refinements, improvements, and replacements can best *rhetorically* present them to others, both to their fellow scientists in the relevant "invisible college" and to broader audiences of lay readers. (See L. Fleck's 1930s classic on the establishment of scientific "facts" for a discussion of the different texts in which innovations are presented to progressively broader audiences [Fleck 1979].)

On this first approach, the *rhetoric* of scientific communication is

3

as distinct from the *logic* of scientific justification as it ever was for Reichenbach or other logical empiricists. Of all the members of the Vienna Circle, Neurath may have been the only one open to a different view. From this first standpoint, it will still be fallacious to assume that a rhetorical analysis of scientific communication has implications about our rational understanding of scientific discovery: one more case of the dread plague of "psychologism" in philosophy of science. "Rhetoric" and "rationality" thus remain polar opposites, and the frontier between psychology and philosophy in the sciences remains an uncrossable iron curtain.

But we have an alternative approach if we use the rhetoric of science to challenge that frontier. In the last resort, can we dissociate the justification of scientific concepts and beliefs entirely from issues of communication in science? Can we talk in real life of "carrying the rational tasks of science through to completion" while ignoring questions about communication? Or do the rational and rhetorical aspects of science legitimately and inescapably overlap? Surely the persuasiveness, and even the "probativeness," of a scientific argument requires us to present it in terms that carry conviction with its audience. Judging the rational merits of a theoretical novelty in, say, electrodynamics, we do well to consider the rhetorical power with which it "speaks to the condition of" the scientists concerned. From this point of view, the rhetoric of communication and the rationality of discovery are no longer separate, as they were before 1960; instead, they now share common ground.

I

When graduate students are admitted as serious scientific collaborators, one of the first things they learn is to match their research results to the expectations of the editor, and reviewers, of the journal to which the manuscripts are submitted. The task may involve issues of *format and style* that differ from journal to journal only mechanically. (In psychology, one can reformat a paper for a new journal at the touch of a computer key.) But it may also involve *substantive* issues. The editorial policies of different journals may reflect different theoretical orientations, which affect what papers they consider seriously, or refuse outright. In such decisions, editors and reviewers place "rational bets" on the direction of long-term theoretical development of their field. Consider the central dogma of molecular biology—that DNA synthesizes RNA, but not vice versa. For a while, some journals would barely consider papers that doubted this assumption, and did not report evidence of reverse transcription. Where such things carry weight with editors, we can no longer separate style from thought.

The problem is to distinguish questions of *style* from questions of *mere style*. One scientist writes more limpid or contorted prose than another, without a difference in their ideas. Or two scientists present their ideas in terms that have different implications without clear evidence supporting one formulation rather than the other. To this extent, *conceptual style* goes far beyond all "literary" considerations, and enters deeply into the fabric of our scientific thinking. Looking back to 1927, for example, we may ask whether Heisenberg's quantum mechanics and Schrödinger's wave mechanics differed other than stylistically: Empirically, the implications of the two theories proved indistinguishable, so they no longer not seem to differ in logically significant respects. Yet we can suggest this only now, in retrospect. At the time, neither philosopher would be happy with this characterization: Each saw the other's argument as misleading and wrongheaded. If our retrospective view is so different, that is because of the water that has flowed under the bridge in the intervening decades. In such a case, we can discover if the differences between two theories are *substantive* as well as *stylistic* only as we go along.

II

Let us consider the distinction between the "rationality" and the "logic" of scientific discovery. This distinction is modern, even parochial: a product of late nineteenth- or early twentieth-century ideas about the philosophy of mathematics and logic, which beg crucial questions. Since Hilbert in mathematics, and Frege and Russell in propositional logic, logical issues have been regarded as *formal*, concerned with, say, consistency, deducibility, and axiomatization. To that extent, matters of "rationality"—concerned with the practical solidity of arguments in, say law, medicine, or science—lie *outside logic*. Only since the 1960s has a concentration on *propositions* given way to an interest in arguments as *utterances*, and the like.

By contrast, in Aristotle's pioneer work on reasoning and communication, the *Organon* (see Aristotle 1938–1960), there was no way to separate logic and rhetoric. For Aristotle, as for modern Greeks, *logos* and *logikos* corresponded to English "reason" and "reasonable" as well as "logical" or "rational." The *Organon* begins with what Aristotle calls *analytics;* but it never *confines* the scope of logic to "analytic" aspects of reason. Instead, it asks on what conditions speech and thought, and communication and action, meet the standards of *to logikon,* including "reasonableness" in all its diversity, along with rationality and formal validity. For instance, the things that make it reasonable to pursue one approach to a science, rather than another, go beyond formal ("analytical") matters: They rest on matters of practical experience or substance. Aristotle's theory of rationality in science, however, is the *Posterior Analytics,* and we may ask why he treats it as analytic at all. To answer that question, we must notice how far his account of natural science, for example, his pre-Darwinian theory of essences, is ahistorical. (His claim that scientific inferences are formal syllogisms holds only if we agree that nature consists of individuals, each possessing the timeless essence of its particular "species.")

Viewing the *Organon* in this way, we have no basis for opposing reason and rhetoric. The manner in which an argument is presented ("Was it convincing, or only heartwarming?"; "Did it hit the point, or go over their heads?") is then as relevant to its intellectual merits as the intrinsic solidity of the case. I am, say, a party to a court case. My attorney presents an intrinsically weak case powerfully, or an intrinsically strong case feebly. Rationally, is the intrinsic force of the case the only relevant point? Or is it not reasonable to criticize the attorney's style of argument as ill adapted to the concerns of the court? Presenting a substantially powerful case in the most persuasive manner—given the character of the judge or jury—is, surely, more "rational" than blowing a case by presenting it feebly, or in terms that give offense.

This contrast between the intrinsic force and extrinsic aspects of an argument requires that we look, also, at the question of probativeness. What gives an argument its rational power to convince? In pre-Cartesian days a distinction was recognized between intrinsic and extrinsic probability. Arguably, this distinction is helpful here as well. The intrinsic probability of a belief is the firsthand knowledge of the matter that inclines us to this belief, regardless of who else agrees or disagrees. Using a phrase from Mill's 1863 *Utilitarianism* (see Mill 1979), it comprises "considerations capable of influencing the intellect to give or withhold assent." And by contrast, the extrinsic probability of a belief is the support lent it by people who speak about the matter with authority (e.g., "learned teachers of the subject generally agree"). How is this distinction relevant to natural science? We never have firsthand knowledge of *all* the considerations relevant to a judgement. For many, we go along with, or trust, the judgement of current experts in the field: "Following Crick the generally accepted account of this matter is . . .", and we introduce new evidence as relevant to problems defined with this consensus as background. My trust in Crick's judgement is, thus, no mere psychological quirk; it is a reasonable judgement about where to locate the current problems in our subject.

III

Keeping in mind the distinctions between formal and substantive reasoning, and between arguments as propositional sequences or human utterances, let us look at the way stations passed in the *Organon* along a spectrum whose extremes are the *Analytics* and the *Rhetoric*. In this way, we can throw light on one central issue for the present discussion, namely, how the purely formal and merely rhetorical aspects of reasoning are related for philosophers working on the natural sciences today.

To begin with issues relevant to conceptual change in science, the manner in which new ideas are added to our ways of thinking is the subject matter of "dialectics." It was never the classical view that dialectics is purely formal, or that its arguments are analytic in any strict sense. At the end of the day, decisions about the directions of conceptual change in the sciences were, and still are, taken in the light of experience, and rest on balancing judgements, weighing a variety of considerations in ways that may involve the use of algorithms (e.g., "significance tests") but are never wholly captured in a single formal algorithm.

A conceptual innovation that yields an empirical treatment of hitherto intractable phenomena in, say, quantum electrodynamics may be accepted as presumably the way to go, despite lacking an underlying theoretical interpretation. Here, the qualification "presumably" marks the conclusion as the outcome of a practical, not a demonstrative, inference. Likewise elsewhere, the balances between predictive power over a limited range of observations, qualitative grasp over a broader range, mathematical simplicity, theoretical intelligibility and the like—where there is no easy way to obtain these all at one and the same time—rest on similar, essential *practical* choices.

Moving on a stage, we reach the topics—both "general" and "special" topics. Now, the subject is the reasoning procedures available in our existing repertory, that is, procedures used in all fields of argument equally (general) and those used only in one, or a few particular fields (special). These patterns of argument—or *topoi*—embraced, at the broadest, general procedures like classifying and arguing from analogy, and, on a narrower, more specific level, local procedures like drawing (optical) ray diagrams, solving (quantum mechanical) eigenequations, and constructing (economic) utility functions. By now, colloquial English has diluted the meaning of the word "topic" so that it is no longer distinct from the phrase "subject"—"What is the topic of today's talk?"—but classically the topic of any talk was the mode of argument put to use in it.

Pausing at other way stations, for Aristotle, a full–scale account of reasoning, thought, and communication finds room also for ethics and politics. Just as we frame arguments and thoughts with an eye to analytic adequacy (e.g., internal consistency), and empirical accuracy (respecting the "grain" of the natural world), so too we frame them with an eye to the interests of the other people with whom we deal, whether we view them as individuals—in the family, as clients, between friends—or as collectives—members of a profession, inhabitants of a local community (*polis*), and so on. Accordingly, Aristotle's approach to logic, reasoning, and rationality—embracing the claims of practice as well as theory, rhetoric as well as substance—left room for no dichotomy of facts and values. For him, "reasonable" arguments would reflect our beliefs about what is good for humans individually or collectively, so there was no division between logical or rational considerations and ethical or political ones. So it is only after moving all along the spectrum (from the analytics, via dialectics, general and special topics, and beyond ethics or politics) that we approach those cases in which we at last face issues of poetics and rhetoric.

Beside the demands of consistency, accuracy, established procedures, and individual or collective good, our arguments are cast in more or less imaginative forms, in varied media, genres, or styles, and "speak to the condition" of their audiences in more or less moving or convincing ways. At no point in the spectrum, from the formal care we take to avoid self-contradictions to our stylistic concern to convey a meaning to the audience, do we cross a clear line that can be used to justify any dichotomy between reason and emotion, or rationality and rhetoric. Rather, in any situation, successful arguments are presented in forms that best responds to all these demands, and one mark of "reasonable" people is the way they match their thoughts, speech, and actions to the entire range of these demands.

IV

As often occurs in philosophy, things that are distinct are not, on that account alone, easy to separate: We can distinguish these aspects of reasoning in theory more easily than we can separate them in practice. The aspects of an argument that convince members of an invisible college, say, are the special topics in which it is framed: The audience is open to conviction because its members regard the types of arguments used as reliable ways to handle such problems. Rhetorically, the best way to pitch an argument so that it "speaks to the condition" of a scientific audience is to cast it in the rational patterns accepted in this scientific subspeciality. Rational force, in this case, serves as the best rhetorical device. More generally, the term "rhetoric" has a broader and a narrower sense:

1. Like early twentieth-century logicians, we may limit the use of "rational" and "logical" to analytic issues and relations. Everything substantial—from dialectics and special topics to poetic choices of genre, and stylistic presentation—must then be included in the rhetorical.

This first broad usage so dilutes the term "rhetorical" that it loses all distinctive force.

2. Alternatively, we may limit our sense of the word "rhetorical" so that it covers only, say, genre, style, vigor, imagery, and audience appeal, and leaves aside all questions of intrinsic force for analysis in rational terms.

This second, narrower sense surrenders half the proper terrain of "rhetoric" to the prejudices of the logical empiricists.

Both senses—the self-diluted and the self-limited—rely on the traditional contrast between rational and rhetorical that we are here seeking to avoid. Can there be a middle way? Certainly in actual practice half a dozen middle ways are possible, and we choose among them as occasion requires. In these middle cases, working at neither end of the spectrum, the rational and the rhetorical are no longer in sharp contrast. Rather, the terms have substantial areas of overlap. The demands of sound reason may be met by the rhetorical force of our intellectual procedures, while the demands of rhetorical force may be met by the rational power of the topics we employ.

V

The point of paying attention, here, to *Organon* and medieval rhetoric was not antiquarian. Aristotle's ideas in philosophy of science are still subject to solid objections; they spring from the pre-Darwinian, ahistorical commitment to *essences* that allowed Aristotle to run his analyses of the deductive syllogism, and scientific explanation, in double harness, lumping them together as branches of the *Analytics*. From our point of view, it is better to dissociate reasoning in science from the formal syllogism, and to discuss the varied method and concept used in dealing with different sets of problems, in different sciences, under the heading of "special topics."

In one respect, the Aristotelian tradition has a genuine advantage in that it reminds us how recently this opposition of reason and rhetoric took hold on the Western mind. In 1600 no such opposition was recognized; by 1700 it was entrenched. Its adoption was a byproduct, less of the Scientific Revolution, than of the success of rationalism in seventeenth-century philosophy. During the 1630s Descartes, in his *Discourse* and *Meditations,* encouraged a demand for Euclidean rigor as the only way to achieve *certainty* in scientific theory: so, he displaced both the fallibilism of the sixteenth-century humanists, and the rhetoricians' concern with utterances as vehicles of knowledge and communication (see Descartes 1980). The resulting shift in the study of reasoning—from utterances to propositions—left on philosophy in Western Europe a mark that would last well into the twentieth century.

The issues raised in this essay were confined to "clearing away the underbrush." It leaves untouched detailed questions about the rhetoric of science, or about scientists as rhetors (i.e., orators). Having lived through the last years of an epoch in philosophy in which reason and rhetoric were at irretrievable odds, so that nothing rational was looked for in rhetoric, and the rationality of arguments was irrelevant to their rhetorical power, I have found the task of clearing away this patch of underbrush and spelling out draft terms for a reconciliation of the two fields a salutary exercise.

REFERENCES

Aristotle. 1938–1960. *Organon,* vols. 1–2. Cambridge, MA: Harvard University.

Descartes, R. 1980. *Discourse on Method and Meditations on First Philosophy.* 1st ed. Indianapolis: Hackett.

Fleck, L. 1979. *Genesis and Development of a Scientific Fact.* Chicago: University of Chicago Press.

Mill, J. S. 1979. *Utilitarianism.* Indianapolis: Hackett.

2

Rhetoric and Rationality in William Harvey's *De Motu Cordis*

Gerald J. Massey
Center for Philosophy of Science, University of Pittsburgh

Howlers in the History of Science and Philosophy

Aquinas and the Eternal World Problem

No reasonable person expects historians of science or philosophy to get every episode or event they happen to investigate completely right—if only for want of time and resources. Still, most people seem to think that historians do get the major events largely right. By contrast, the Massey-Hintikka-Newman Law states that in science and philosophy any historical episode that can be misunderstood will be misunderstood to a degree directly proportional to its importance (J. Hintikka and R. Newman are two of the many scholars who have endorsed this law). This law follows as a quasi-corollary from Murphy's Law, which declares that whatever can go wrong will go wrong. Empirically, what makes the Massey-Hintikka-Newman Law compelling are the numerous case studies that show how badly professional historians, and scientists and philosophers as well, have gone wrong in relating key episodes in the history of science and philosophy. For present purposes, one or two illustrations will suffice.

The most sizzling philosophical-theological topic in the thirteenth century Christian West was the so-called eternal world problem. The newly introduced corpus of Aristotle's writings bristled with seemingly cogent arguments for the eternity of the world (universe). Enhanced by the prodigious authority accorded to the Peripatetic, these impressive arguments threatened to undermine the magisterium of the Christian Church which had long insisted on temporal creation (creation in time) as part of the divinely revealed deposit of faith. Perhaps the most crucial part of Aquinas's mission to reconcile Aristotle with Christian teaching, therefore, centered on the eternal world problem.

The task was not easy, for Aquinas had to steer a hazardous course between Scylla and Charybdis. The rock was the Christian doctrine that the world was created in time, that is, that there was a first moment of time. (Creation in time is often confused with creation *ex nihilo*. Aquinas [1952, Ia, q. 44, a.2; 1923, vol. 2, chap. 16] held that the doctrine of creation *ex nihilo* could be demonstrated within philosophy.) The whirlpool was the august status that Aquinas attributed to this teaching, namely, the stature of a Dogma, that is, a revealed truth whose verification surpasses human capacities. (Interestingly, many of Aquinas's Franciscan contemporaries, notably Bonaventure [Van Veldhuijsen 1990, 24–28; Davidson 1987, 140] classified the doctrine of temporal creation as one of the prolegomena to faith, i.e., as a revealed proposition that can be demonstrated or proven within philosophy.) Thus, while a demonstration that the world was created in time might be welcomed by some Christian thinkers, for Aquinas such a proof would clash with the doctrine's lofty status as a Dogma. A demonstration that the world is eternal (time has no first moment) was totally anathema, of course. But because such a demonstration would contradict revealed "truth," Aquinas never worried about someone's finding one.

So how did Aquinas navigate these dangerous waters? The course he laid between Scylla and Charybdis consists of a metademonstration that there is no demonstration for either the temporality or the eternity of the world (1952, Ia, q. 46, a. 2; and 1923, vol. 2, chaps. 32–38), an achievement which virtually all neoscholastic philosophers and historians have applauded, many trumpeting it as one of Aquinas's greatest achievements.

But the historical tale has already gone awry. Historians and philosophers have failed to notice that Aquinas offered a metademonstration only that there is no demonstration *propter quid* for temporal or eternal creation. The question remained whether there might be a demonstration *quia* for one of the two alternatives. (Roughly speaking, demonstrations *propter quid* are explanatory proofs that move from causes to their effects, whereas demonstrations *quia* are nonexplanatory proofs that move from effects to their causes.) Aquinas ought not to have concluded, therefore, that the doctrine of temporal creation surpasses human reason. His metademonstration entitled him to conclude that temporal creation surpasses human reason so long as there are no demonstrations *quia* of either the temporality or the eternity of the world. In the parallel matter of God's existence, Aquinas properly did not infer from his metademonstration that there are no demonstrations *propter quid* of the existence of God that the issue transcends human reason (1952, Ia, q. 2. a. 2; and 1923, vol. 1, chap. 12). On the contrary, he searched for and found numerous demonstrations *quia* that God exists, among them his celebrated Five Ways (1952, Ia, q. 2, a. 3; and 1923, vol. 1, chap. 13).

Aquinas's metademonstration of the indemonstrability of the world's temporality or eternity secures for temporal creation the status of a Dogma only if it is supplemented by a demonstration that there are no *quia* demonstrations of either disjunct. But between Aquinas and such a supplemental metademonstration lay an Arabian sandbar that stretched from the rock to the whirlpool. Aquinas took due notice of this navigational hazard only at the end of his life, just in time to see his ship founder on it. This deadly menace was Al-Ghazali's ingenious proof of temporal creation.

Al-Ghazali (his Latin name was Algazel) established the temporality of the world by propounding a reductio proof that showed the untenability of the eternal world thesis (see Davidson 1987, 123–124). Following are the several premises of Al-Ghazali's slightly reconstructed reductio proof:

1. The world is eternal (i.e., there is no first moment of time).
2. Humans have always lived in the world.
3. Humans have always reproduced at or above some modest rate, for example, one birth per century.
4. The human soul is immortal.
5. The human soul is individual.
6. The human soul retains its individuality after death.
7. Actual infinite multitudes are impossible. (An actual infinite multitude is an infinite set of things all of which coexist at some moment or other.)

Al-Ghazali posed the following question: If his seven premises are all true, how many souls are now in existence? The answer is immediate: an actual infinite multitude of souls—contra the seventh premise. This contradiction shows that at least one of the premises must be rejected. Al-Ghazali believed that Christian thinkers would be compelled to reject the first premise, the eternity of the world. In this supposition, Al-Ghazali was correct. The impossibility of actual infinite multitudes was virtually a philosophical dogma in the thirteenth century; Aquinas believed that he had *demonstrated* its truth (1952, Ia, q. 7, a. 4). Additionally, religious faith committed Christian thinkers to premises two through six. Take premise two, for example. Believing that the world was created for humans, it made no sense to Christians that the world might lie fallow for an eternity before God created human beings. Therefore, Al-Ghazali had apparently found a demonstration *quia* for the temporality of the world, thereby undermining the status Aquinas accorded to this doctrine as a Christian Dogma surpassing human reason.

How, then, did Aquinas deal with Al-Ghazali's demonstration? Although Aquinas refers to it in many places as a difficult argument, his practice is nevertheless to dismiss it as a weak argument because it has many premises (1952, q. 46, a. 2; and 1923, vol. 2, chap. 38). This dismissal is not only superficial but disingenuous. Aquinas fails to mention that his philosophy and theology together commit him to the necessity of all of Al-Ghazali's premises save the first. Most disingenuously, Aquinas does not reply in his own person to Al-Ghazali's argument, but hides behind a smoke screen of others' views. For example, Aquinas (1952, Ia, q. 46, a. 2, *ad* 8; 1923 vol. 2, chap. 38) responds to Al-Ghazali's argument by canvassing how some thinkers who posit an eternal world have responded to Al-Ghazali's purported demonstration. Aquinas fails to point out that none of these responses can be espoused by a Christian-Aristotelian like himself. Only sophistry, then, keeps Aquinas from acknowledging Al-Ghazali's argument as a bona fide *quia* demonstration of the temporality of the world. But in view of the status Aquinas accorded to the temporal-world doctrine as a Christian Dogma, it is no surprise that he unfairly and disingenuously dismissed Al-Ghazali's menacing argument.

Only in a tract written near the end of his life does Aquinas (1927) come close to according Al-Ghazali's argument the attention and analysis it merits. The tract contains a discussion of error concerning the power of God. Aquinas notes that thinkers can err either by exaggerating God's power or by minimizing it. Although exaggeration and minimization both lead to false claims about God, Aquinas calls attention to a crucial difference between the two: Those who have exaggerated God's power have been deemed pious by the Christian Church, while those who have minimized it have been branded heretics. Given these musings, one can readily predict what Aquinas will do in the tract when he emerges from behind his customary smoke screen.

To get around Al-Ghazali's argument, Aquinas first offers several rejoinders that would defuse the argument if he were able to endorse them, namely, that God could have created the world without human beings, or that God could have created the world-without-humans from eternity and then created humans at some finite time in the past. Aquinas neglects to mention that both responses are closed to him as a Christian theologian committed to the proposition that God created the world for humans. Then, at long last, Aquinas proposes his own way out of Al-Ghazali's argument: He exaggerates God's power by claiming that no one has proven that God cannot create actual infinite multitudes—something he purports to prove to be impossible (1952, Ia, q. 7, a. 4). This maneuver enables Aquinas to reject premise seven rather than premise one, thereby defusing Al-Ghazali's reductio argument.

The treatment accorded Al-Ghazali's argument and the claim about God's ability to create actual infinite multitudes—hence, to be able to do impossible things—differ so radically from anything Aquinas does or says elsewhere that one might be tempted to question Aquinas's authorship—despite the fact that this tract is mentioned in the *Correctoria Corruptorii,* a collection of spirited defenses of Aquinas by fellow Dominicans against the scathing attacks of the Franciscan Friar William de la Mare's *Correctorium Fratris Thomae,* all of which were written within a few years of Aquinas's death (Hoenen 1990, 39–40). But in fact Aquinas (1927) says nothing that one would not expect him to say *in a context where he accords Al-Ghazali's argument the serious treatment it merits.* Even the way he warps philosophical dogmas (the impossibility of actual infinite multitudes, the equivalence of logical possibility with God's power) to bring them into line with his lofty conception of the doctrinal status of temporal creation illustrates one of his own methodological maxims, to wit: In the event of conflict, Christian thinkers should adapt their philosophical views to

Christian teaching because incompatibility with such teaching is the best evidence of philosophical error available to humans (1923, vol. 1, chap. 7).

Coincidence of the Rhetorical and Rational Dimensions

Subsumption of the Massey-Hintikka-Newman Law under Murphy's Law hardly explains why historians of science or philosophy typically mangle the momentous episodes they tackle. But what does explain these blunders is failure to recognize that in scientific and philosophical argumentation the rhetorical and the rational blend together. As S. Toulmin writes in "Science and the Many Faces of Rhetoric," "The demands of sound reason may be met by the rhetorical force of our intellectual procedures, while the demands of rhetorical force may be met by the rational power of the topics [patterns of argument] we employ." Because the rhetorical and rational dimensions largely coincide, rationality demands that scientists and philosophers present their arguments and reasonings in terms and manners crafted to carry conviction with their audiences. That is, rationality requires them to speak to the condition of their audiences, a point Toulmin also stresses.

Galen (1952) propounded the same idea when reproaching Erasistratus for the latter's failure to come to grips with Hippocrates's theory of the attractive power of the stomach in deglutition:

"My good sir, do not run us down in this rhetorical fashion without some proof; state some definite objection to our view, in order that either you may convince us by a brilliant refutation of the ancient doctrine, or that, on the other hand, we may convert you from your ignorance." Yet why do I say "rhetorical"? For we too are not to suppose that when certain rhetoricians pour ridicule upon that which they are quite incapable of refuting, without any attempt at argument, their words are really thereby constituted rhetoric. *For rhetoric proceeds by persuasive reasoning: words without reasoning are buffoonery rather than rhetoric.* Therefore, the reply of Erasistratus in his treatise "On Deglutition" was neither rhetoric nor logic. (P. 180–81)

Attention to rhetorical context, therefore, is the key to a sound appreciation of major episodes in the history of science and philosophy. We saw, for example, how attention to the rhetoric of Aquinas's (1927) tract reveals not only what the author is going to do in the tract but also why. The same opusculum contains caustic criticism of some contemporary thinkers who, snarls Aquinas, seem to think that philosophy began with them. Had Galen been alive to read these abusive lines, he would instantly have interpreted this rhetorical clue to mean that Aquinas was acutely uncomfortable with the position Al-Ghazali's argument had forced him to take in the tract. Similarly, the superficial dismissals of Al-Ghazali's argument as weak, frequently found in Aquinas's other writings, testify that Aquinas is either unable or unwilling to deal seriously with this threat to the magisterium of the Christian Church.

Abusive argumentation distressed William Harvey (1578–1657) so much that he became semireclusive. Concerning Harvey's attitude toward such argumentation, his friend G. Ent (1952, 330) related the following: "[He] is wont to say that it is argument of an indifferent cause when it is contended for with violence and distemper; and that truth scarce wants an advocate." Like Galen, Harvey would have had no trouble deciphering the sarcasm Aquinas (1927) directed at certain of his contemporaries. Nor would Harvey have been slow to see that Aquinas's (1923, vol. 1, chap. 6) denunciation of Mohammed and Islam was prompted by the fact that most of the charges Aquinas levels against them apply mutatis mutandis to Christ and Christianity.

Aquinas (1927) does not explain why he immunizes premises four through six against falsification by Al-Ghazali's reductio argument. Does this omission constitute an oversight or slipshod analysis? It would so appear only to someone ignorant of the condition of Aquinas's audience. Far from being an argumentative defect, this omission illustrates Aquinas's sensitivity to his audience. The theologians and philosophers for whom the tract was intended knew as well as did Aquinas that the Christian Church looked upon these premises as necessary propositions included among the revealed truths. Aquinas had no need to point this out or to explain why he did not take Al-Ghazali's reductio argument to discredit any of these religiously privileged premises. So, we see again how rationality and rhetoric coincide, and how failure to appreciate the one entails failure to appreciate the other. Let us now examine how failure to appreciate the overlap of these two dimensions has distorted one of the greatest episodes in the history of medical science, namely, Harvey's discovery of the circulation of the blood, and how attention to the overlap illumines it.

Harvey as Experimenter: The Myth and the Mystery

The Myth: Harvey, the Bungling Experimenter

In a well-known and influential article, F. Kilgour (1987) character-izes Harvey—the physician hailed as the father of modern medicine—as an experimental klutz. Kilgour tries to reconcile Harvey's alleged experimental bungling with his manifest genius by underscoring the novelty of Harvey's quantitative approach to biology, "Harvey was the first biologist to use quantitative methods to demonstrate an im-portant discovery. To weigh, to measure, to count and thus to arrive at truth was such a new idea in the 17th century that even a man of Harvey's genius could do it only badly" (ibid., 11).

On its face, Kilgour's claim is so improbable as to be incredible. After all, Harvey painstakingly dissected more than 80 different kinds of animals and reported on this anatomical work with uncommon care, precision, and insight (Anonymous 1952, 263). And no one can read Harvey's *Exercitatio Anatomica de Motu Cordis et Sanguinis in Animalibus* (1628; hereafter, *De Motu Cordis*) without being im-pressed by Harvey's intelligence, the remarkable skill and ingenuity he brought to anatomy, his commitment to careful observation as the ultimate arbiter of anatomical and biological claims, and the emphasis he placed on measurement and computation in the study of the mo-tions of the heart and blood in animals. Moreover, in chapter 9—argu-ably the central chapter of the book—Harvey repeatedly appeals to calculations concerning the quantity of blood coursing through the heart, arteries, and veins, clearly suggesting that he had made the asso-ciated measurements with his customary thoroughness and precision. Furthermore, Harvey spent more than nine years perfecting his theory, lecturing on it, demonstrating it to friends and colleagues, and replying to their objections before committing it to print (Harvey 1952a, 267). Surely, at least some of his friends and colleagues could weigh and measure and count, or at least take an accurate pulse, even if Harvey could not! Could such a premier scientist really have been a bungler as an experimenter?

Kilgour, however, has four principal reasons for labeling Harvey an experimental klutz. First, Harvey (allegedly) grossly underestimated human cardiac output. By "cardiac output" Kilgour means the aver-age amount of blood (by weight) ejected from the left ventricle of the heart into the arteries during systole in an adult. Today "cardiac output" means the volume of blood ejected by the heart in one minute (Berkow et al. 1987, 378). And "stroke volume" means the quantity of blood ejected per heartbeat (Fraser, et al. 1991, 8). Thus, Kilgour's concept of "cardiac output" corresponds closely to stroke volume, but not to the modern concept of cardiac output. To avoid misunderstandings, let us introduce the term "stroke weight" for what Kilgour calls "cardiac output," meaning the amount of blood (by weight, not volume) ejected per heartbeat. Moreover, because both Kilgour and Harvey are concerned with the amount of blood *by weight* ejected from the left ventricle over a specified time period, we will adapt modern usage slightly to facilitate discussion across centuries. Let *cardiac output** be the *weight* of blood ejected by the heart (left ventricle) in one minute and let *half-hour cardiac output** be cardiac output* multiplied by 30. (The asterisks are meant to suggest weight as opposed to volume of blood per time interval.) Concerning Harvey's estimate of stroke weight, Kilgour comments, "The measurement of cardiac output [stroke weight] is a tough problem, and even today there are wide variations in the measurements obtained by the various methods. But Harvey got a figure that is only *one-eighteenth of the lowest estimate used today*. How could he have arrived at such a ridiculously incorrect figure and at the same time have used it successfully to demonstrate such an important discovery?" (1987, 14; emphasis added).

Kilgour's second reason concerns Harvey's alleged gross underestimation of the stroke weight of sheep:

Harvey can certainly be excused for not obtaining any close estimate of the output [stroke weight] of the human heart, but he got virtually as poor results when he tried to measure the output [stroke weight] of a sheep's heart. If he had severed a sheep's aorta and weighed the amount of blood ejected in one minute while counting the heartbeats for that minute, he could have obtained a reasonably accurate figure for cardiac output [stroke weight] in sheep. But he never performed that obvious experiment. (Ibid.)

Kilgour's third reason is Harvey's alleged gross underestimation of the average adult pulse rate, recognized today to be about 72 beats per minute (Considine 1983, 1462), "Harvey also missed the mark widely in his other important measurement—the pulse rate. Somehow he counted it to be 33 per minute, about half the actual average rate, and although he obtained other values, he generally used this figure. We cannot explain this error on the ground that it is a difficult measurement to make; *why he went so wrong will always be a mystery*" (Kilgour 1987, 14; emphasis added).

Kilgour's fourth reason follows from his second and third allegations, namely, Harvey's absurdly low estimate of the rate of blood flow, "With his two estimates—3.9 grams for cardiac output [stroke weight] and 33 pulse beats per minute—[Harvey] obtained a figure for the rate of blood flow [cardiac output*] which is *one-36th of the lowest value accepted today*" (ibid.; emphasis added).

The alleged grossly inaccurate data used by Harvey relate to his demonstration of the first of three theses or propositions propounded in the first paragraph of chapter 9 of *De Motu Cordis*, namely, that the heart pumps blood into the arteries in quantities so large that it could not possibly be supplied by ingesta and so swiftly that the entire mass of blood passes through the heart in a short time. While conceding that Harvey's demonstration of the first proposition was successful despite being allegedly marred by grossly inaccurate data, Kilgour thinks that Harvey did an even less competent job of demonstrating his second proposition than he did his first. "Less impressive was Harvey's demonstration of the second proposition: that the amount of blood going to the extremities is much greater than is needed for the nutrition of the body. He used no specific measurements and argued largely by inference" (ibid.).

Kilgour's briefs against Harvey have no force whatsoever. Only Kilgour's lamentable neglect of Harvey's rhetorical context lends even superficial plausibility to his accusations.

Harvey's Rhetorical Context

Despite Harvey's alleged use of grossly inaccurate data, even Kilgour recognizes that Harvey *succeeded* in demonstrating his main point, namely, that the Galenic theory of the heart and blood was bankrupt, "Even with his faulty calculations, Harvey proved his main point: that each half-hour the blood pumped by the heart far exceeds the total weight of blood in the body. This was a blow to the Galenic concept, for it was obvious that the food a man eats could not produce blood continuously in any such volume" (ibid.). In the foregoing passage Kilgour unwittingly calls attention to one of Harvey's two principal objectives in *De Motu Cordis*, namely, to liberate anatomists and physicians from the Galenic theory so that they might be free to consider and accept his own rival theory (the other principal objective), which posited a continual circulation of the blood in contrast to the continual Galenic transformation of food to replenish the blood sent to nourish the extremities.

Harvey was keenly aware of his rhetorical context. He anticipated that his novel theory would meet with bitter opposition fueled by long-entrenched theory, habit, tradition, and conservatism, "But what remains to be said upon the quantity and source of the blood which thus passes [from the veins, through the heart and lungs, into the arteries] is of so novel and unheard-of character, that I not only fear injury to myself from the envy of a few, but I tremble lest I have mankind at large for my enemies, so much doth wont and custom, that become as another nature, and doctrine once sown and that hath struck deep root, and respect for antiquity influence all men: still the die is cast, and my trust is in my love of truth, and the candour that inheres in cultivated minds" (Harvey 1952a, chap. 8, 285).

That Harvey consciously took into consideration the condition of his audience is evident. First he usually employs the time period of one-half hour when calculating blood flow. As his contemporaries would have been aware, Galen had made one-half hour particularly germane to Harvey's objectives, "[T]he great artery need not be divided, but a very small branch only (*as Galen even proves in regard to man*), to have the whole of the blood in the body, as well that of the veins as of the arteries, drained away in the course of no long time—*some half hour or less*" (ibid., chap. 9, 287; emphasis added). Similarly, in the introduction to *De Motu Cordis*, Harvey painstakingly refutes the doctrine that the arteries contain only air, "That it is blood and blood alone which is contained in the arteries is made manifest *by the experiment of Galen,* by arteriotomy, and by wounds; for from a single artery divided, *as Galen himself affirms in more than one place,* the whole of the blood may be withdrawn in the course of *half an hour, or less.* (ibid., 269–70; emphases added). Galen had simply left Harvey no alternative. Only by selecting the time period of one-half hour for calculations of blood flow could Harvey speak to the condition of his seventeenth-century audience.

Second, Harvey uses a well-chosen set of assumed or hypothetical values for human and animal pulse rates when making calculations to determine half-hour cardiac output*. As Kilgour notes, the slowest underlying pulse rate Harvey uses in chapter 9 for humans is 33 beats per minute. When rounded off—Harvey's common practice—33 beats per minute yields the round figure of 1,000 beats *per half hour,* thereby simplifying the calculations of half-hour cardiac output* for physicians unaccustomed to quantitative reasoning. Indeed, in chapter 9, Harvey gives illustrative calculations of half-hour cardiac output* based upon four different underlying pulse rates—namely, his minimum pulse rate of 33 beats per minute together with three other pulse rates that are respectively twice, thrice, and four times this minimum. That is, Harvey deploys pulse rates that yield 1,000; 2,000; 3,000; and 4,000 beats per half hour, respectively (when rounded off), thereby simplifying calculations of half-hour cardiac output*. Moreover, Harvey's strategy to discredit the rival Galenic system requires him to employ patently conservative values of pulse rate and stroke weight. But 33 beats per minute is the only possible value for pulse rate that is both patently conservative and possesses the aforementioned virtue of simplifying calculations of half-hour cardiac output*. Thus, the choice of 33 beats per minute for his illustrative minimum pulse rate was forced upon Harvey by the condition of his audience and his rhetorical context. Had Harvey employed his second slowest pulse rate, [66 beats per minute] as the minimum in his human blood-flow calculations—undoubtedly closer to what he believed to be the average adult pulse rate—Harvey would have been less convincing in constructing an unanswerable case against the Galenic theory, one of the two principal objectives of *De Motu Cordis.*

Like rhetoric and rationality, the rhetorical and scientific backgrounds of scientific discoveries largely coincide. It will be helpful, then, to review Kilgour's concise formulation of the Galenic theory Harvey attacked:

According to Galen, chyle (a kind of lymph) passed from the intestines to the liver, which converted it to venous blood and at the same time added a "natural spirit." The liver then distributed this blood through the venous system, including the right ventricle of the heart. Galen knew from experiment that when he severed a large vein or artery in an animal, blood would drain off from both the veins and the arteries. He realized, therefore, that there must be some connection between the veins and the arteries, and he believed he had found such a connection in the form of pores in the wall dividing the left side of the heart from the right. He argued that the venous blood oozed through these supposed pores to the left heart, was there charged with *vital spirit* coming from the lungs and thus took on the bright crimson color of arterial blood. According to the Galenic scheme, blood flowed to various parts of the body through both the veins and the arteries to supply the body members with nourishment and spirit. There was no real circulation or motive power; the blood in the vessels simply ebbed back occasionally to the heart and lungs for the removal of impurities. (1987, 12–13)

Two decades before Harvey's birth, Vesalius (1514–1564) announced that the pores in the septum of the heart which Galen claimed to have found simply did not exist. Then Realdo Colombo, Vesalius's successor in anatomy at Padua, discovered that blood flows from the right ventricle via the pulmonary arteries into the lungs and thence via the pulmonary veins into the left auricle. Important as these discoveries were to Harvey's work, the discovery that unquestionably contributed most was made by Harvey's teacher at Padua, Fabricius of Aquapendente. Fabricius had found little doors (valves) in the veins, whose function Fabricius took to be to retard the flow of blood to the extremities. Harvey told Robert Boyle that reflection on the purpose of these valves had led him to the theory of the circulation of the blood (ibid., 11, 13). So great was his admiration and respect for Fabricius that Harvey states in the preface to *De Motu Cordis* that he does not think it "right or proper" (1952a, 268) to dispute with gifted anatomists who have been his teachers. Pertinently, Harvey then remarks that "Fabricius of Aquapendente, although he has accurately and learnedly delineated almost every one of the several parts of animals in a special work, has left the heart alone untouched" (ibid., 274), as if Fabricius had thereby bequeathed the investigation of this organ to other anatomists, Harvey among them.

Harvey's Rhetorical Genius: An Overview

Harvey's theory of the heart and blood rests solidly upon six pillars. He succinctly highlights three, "[It] is imperative on us first to state what has been thought of these things by others in their writings, and what has been held by the vulgar and by tradition, in order that what is true may be confirmed, and what is false is set right by *dissection, multiplied experience,* and *accurate observation*" (ibid.; emphasis added).

A fourth major pillar of Harvey's work is *comparative anatomy,* "[It] will be proper . . . to contemplate the motion of the heart and arteries, not only in man, but in all animals that have hearts" (ibid., 273). For example, concerning the widespread ignorance of the heart-lungs-blood complex, Harvey remarks that "they plainly do amiss who, pretending to speak of the parts of animals generally, as anatomists for the most part do, confine their researches to the human body alone, and that when it is dead. . . . Had anatomists only been as conversant with the dissection of the lower animals as they are with that of the human body, the matters that have hitherto kept them in a perplexity of doubt would, in my opinion, have met them freed from every kind of difficulty" (ibid., chap. 6, 280). But as Harvey (1952c) relates, some opponents flayed him for the anatomical use he made of lower animals, "There are some, too, who say that I have shown a vainglorious love of vivisections, and who scoff at and deride the introduction of frogs and serpents, flies, and others of the lower animals upon the scene, as a piece of puerile levity, not even refraining from opprobrious epithets" (p. 313).

Though he took it to new heights, comparative anatomy did not originate with Harvey. Galen, too, had championed comparative anatomy, for example, in the investigation of the nature and functions of large hollow human organs such as the stomach and uterus:

What prevents us, then, from taking up these first and considering their activities, conducting the enquiry on our own persons in regard to those activities which are obvious without dissection, and, in the case of those which are more obscure, dissecting animals which are near to man; not that even animals unlike him will not show, in a general way, the faculty in question, but because in this manner we may find out at once what is common to all and what is peculiar to ourselves, and so may become more resourceful in the diagnosis and treatment of disease. (1952, 200)

Embryology, to which Harvey made important contributions, forms a fifth pillar of his work. Harvey often bolsters or illustrates his points about the heart and blood (and related matters such as the lungs) by reference to the development of animal embryos and sometimes even of the human fetus (e.g., see 1952a, chap. 6).

The sixth pillar of Harvey's work consists of *appeal to high authorities* such as Aristotle and Galen, both of whom Harvey frequently cites for support. For example, in chapter 7 of *De Motu Cordis*, Harvey appeals to Galen—not hesitating to put a bit of his own spin on the views of "that great man, that father of physicians"—in support of Colombo's recently advanced theory that blood passes from the veins through the lungs to the aorta via the auricles and ventricles of the heart. In his tract *The First Anatomical Disquisition on the Circulation of the Blood, Addressed to John Riolan*—an anatomist who had tried to parry Harvey's evidence and arguments by grafting a partial circulation of blood onto the Galenic system—Harvey boasts that his own theory better confirms the doctrines of ancient medicine, "The circulation of the blood does not shake, but much rather confirms the ancient medicine; though it runs counter to the physiology of physicians, and their speculations upon natural subjects, and opposes the anatomical doctrine of the use and action of the heart and lungs, and rest of the viscera" (1952d, 306).

Should Harvey's invocation of ancient authorities be characterized as hypocritical and self-serving rhetoric—he did, after all, profess "to learn and to teach anatomy, not from books but from dissections; not from the positions of philosophers but from the fabric of nature" (1952a, 270). But there is nothing disingenuous or irrational about appeals to authority even when made by empirically-minded scientists so long as they make observation and experience the ultimate grounds of assent or dissent. And this Harvey surely did.

In addition to the Galenic theory, Harvey faced some internal problems. First, his theory harbored a gap, namely, the unobserved and unknown routes by which blood finds its way from the arteries to the veins in the extremities. Harvey is especially sensitive to the reluctance of anatomists to accept claims about blood flow unless accompanied by direct evidence of the routes taken by the blood. For example, Harvey acknowledges that even Galen might have hesitated to accept the thesis that blood flows from the veins via the heart into the arteries "like all who have come after him, even to the present hour, because he did not perceive the route by which the blood was transferred from the veins to the arteries, in consequence . . . of the intimate connexion between the heart and the lungs." (ibid., chap. 5, 280).

Apropos this missing link in Harvey's theory, Kilgour says, "Thirty-three years after the appearance of *De Motu Cordis* the Italian anatomist Marcello Malpighi filled in that link by discovering the capillaries and so completed Harvey's scheme" (1987, 15). But Harvey was already several years dead and his theory long accepted by the time Malpighi discovered the capillaries. How, then, did Harvey extricate his theory from this grave internal predicament? Harvey never found the elusive arteries-to-veins passages, which from his work on vascular valves he knew must exist. Harvey's best guess in *De Motu Cordis* was this lame tripartite disjunctive hypothesis, "I have here to cite certain experiments [concerning the role and function of the valves in the vein] from which it seems obvious that . . . in the limbs and extreme parts of the body the blood passes either immediately by anastomosis from the arteries into the veins, or mediately by the pores of the flesh, or in both ways, as has already been said in speaking of the passage of the blood through the lungs" (1952a, chap. 11, 289).

Later in the same chapter, Harvey reiterates the foregoing disjunctive claim. But Harvey later drops the anastomoses disjunct altogether and embraces the pores-or-interstices disjunct (see Elkana and Goodfield 1968 for an analysis), "These then, are, as it were, the very elements and indications of the passage and circulation of the blood, *viz.*, from the right auricle into the right ventricle; from the right ventricle by the way of the lungs into the left auricle; thence into the left ventricle and aorta; *whence by the arteries at large through the pores or interstices of the tissues into the veins,* and by the veins back again with great rapidity to the base of the heart" (Harvey 1952c, 326; emphasis added).

How should a theorist argue for a novel theory that displays an embarrassing hole? Rather than fret over the gap in his theory, Harvey chose to bludgeon the Galenic theory by emphasizing the cardinal fact that had convinced him of the circulation of the blood, namely, the huge quantity of blood that passes rapidly and unidirectionally from the veins via the heart and lungs into the arteries, a quantity much too large for the Galenic theory even to begin to cope with:

I frequently and seriously bethought me, and long revolved in my mind, what might be the quantity of blood which was transmitted, in how short a time its passage might be effected, and the like; and not finding it possible that this could be supplied by the juices of the ingested aliment [as the Galenic theory demands] without the veins on the one hand becoming drained, and the arteries on the other getting ruptured through the excessive charge of blood, unless the blood should somehow find its way from the arteries into the veins, and so return to the right side of the heart: I began to think whether there might not be a MOTION, AS IT WERE, IN A CIRCLE. (1952a, chap. 8, 285)

How Harvey carried out this quantitative assault on the Galenic theory is explained in the next section.

Had Harvey attempted to hide or cover up the hole in his theory, opponents would have quickly found it and discredited him. So, by forthrightly admitting the defect, he gained some credibility. This credibility was enhanced by his habit of volunteering evidence unfavorable to his views. For example, concerning his belief that "first the auricles *contract,* and then subsequently the heart (the ventricles) itself *contracts,*" (ibid., chap. 4, 277), Harvey comments, "But I think it right to describe what I have observed of an opposite character: the heart of an eel, of several fishes, and even of some animals taken out of the body, beats without auricles; nay, if it be cut in pieces the several parts may still be seen contracting and relaxing; so that in these creatures the body of the heart may be seen pulsating, palpitating, after the cessation of all motion in the auricle" (ibid.).

Only somewhat less vexing than the missing arteries-to-veins routes was Harvey's failure to accomplish the major objective of chapter 13 of *De Motu Cordis,* "[To] explain . . . in what manner the blood finds its way back to the heart from the extremities by the veins, and how and in what way these are the only vessels that convey the blood from the external to the central parts" (ibid., chap. 13, 293). Harvey supplies no such explanation in chapter 13, but contents himself with describing the function of the venous valves. When the deferred explanation finally makes its appearance in chapter 15, it is a disappointment. Harvey waxes almost mystical in speaking about the blood's spontaneous inclination to flow toward the heart (the center) and its natural disposition to aggregate at its source. He recognizes that blood "is forced from the capillary veins into the smaller ramifications, and from these into the larger trunks by the motion of the extremities and the compression of the muscles generally" (ibid., chap. 15, 297). But, without appeal to pressure gradients, (a development that lay well in the future) Harvey could mount only a weak case for return blood flow. He did wisely, therefore, not to call overmuch attention to the matter.

Ever inclined to the offensive, Harvey closes *De Motu Cordis* with a rhetorical flourish. He challenges opponents to explain the anatomical structures and physiological phenomena accounted for by his theory without invoking the circulation of the blood, "for it would be very difficult to explain in any other way to what purpose all is constructed and arranged as we have seen it to be" (ibid., chap. 17, 304).

The Rhetorical and Scientific Genius of Harvey's Blood Flow Calculations

Harvey's rhetorical genius shines forth in his calculations of hypothetical blood flow in humans and animals, the very calculations in which he employs the alleged grossly inaccurate values of pulse rate and stroke weight that moved Kilgour to characterize Harvey as an experimental klutz. *Harvey's rhetorical strategy consists in showing that even extremely conservative assumptions about pulse rate and stroke weight lead to quantities of blood flow so large as to overwhelm the Galenic theory.* The values Harvey uses for pulse rate and stroke weight are *not* chosen for their accuracy—as Kilgour's analysis presupposes—but for their evident conservatism. It is to be expected, then, that these values will fall considerably short of what today are the accepted values. Harvey would deserve criticism for using these values only if they had been proposed as accurate or realistic, which they decidedly were not.

The bulk of Harvey's blood-flow calculations are reported or suggested in chapter 9 of *De Motu Cordis*. Note that throughout the text Harvey uses the units of weight of the Apothecary System, not those of the Customary English System. To avoid confusion, we will follow Harvey's example. The relevant Apothecary units along with their metric equivalents, compiled from Morris (1969, 811–13), are as follows:

Apothecary Units of Weights

---

1 *grain* = 0.0648 grams
1 *scruple* = 20 grains = 1.296 grams
1 *dram* = 3 scruples = 60 grains = 3.888 grams
1 *oz* = 8 drams = 24 scruples = 480 grains = 31.104 grams
1 *lb* = 12 oz = 96 drams = 288 scruples = 5,760 grains = 373.248 grams

Early in chapter 9, Harvey asks the reader to "assume, either arbitrarily or from experiment, the quantity of blood which the left ventricle of the heart will contain when distended to be, say two ounces, three ounces, one ounce and a half—in the dead body I have found it to hold upwards of two ounces" (1952a, 286). That is, Harvey asks the reader to entertain three different values of left-ventricle capacity (when dilated). Then, he asks the reader to "suppose as approaching the truth [*verisimili coniectura ponere liceat*] that the fourth, or fifth, or sixth, or even but the eighth part of its charge is thrown into the artery at each contraction" (ibid.). That is, Harvey asks the reader to entertain four conjectures about the so-called *eject fraction*, that is, the fraction of blood ejected from the left ventricle during systole. (For a person at rest, the eject fraction is today estimated at about 3/5 [Fraser et al. 1991, 9].) Then, by multiplying assumed ventricle capacity (the amount of blood that the left ventricle has been assumed to contain) by assumed eject fraction (the fraction of blood that has been assumed to be ejected upon contraction), Harvey arrives at the following partial list of values for stroke weight, "[T]his would give either half an ounce [4 drachms or "drams"], or three drachms, or one drachm of blood as propelled by the heart at each pulse into the aorta" (1952a, chap. 9, 286). All that Harvey now needs to complete his blood flow calculations are the number of heartbeats in one-half hour, which he posits thus, "Now in the course of half an hour, the heart will have made more than one thousand beats, in some as many as two, three, and even four thousand" (ibid.). In other words, Harvey's hypothetical calculations are based upon four different pulse rates, yielding 1,000; 2,000; 3,000; and 4,000 beats in one-half hour. Harvey proceeds to calculate hypothetical half-hour cardiac output* by multiplying the assumed stroke weight by the assumed number of beats in one-half hour, and then records the following partial list of values of half-hour cardiac output*, "Multiplying the number of drachms propelled by the number of pulses, we shall have either one thousand half ounces [4,000 drams], or one thousand times three drachms, or a like proportional quantity of blood, according to the amount which we assume as propelled with each stroke of the heart, sent from this organ into the artery" (ibid.).

As can be readily seen from table 2.1, Harvey's various assumptions generate forty-eight different values of half-hour cardiac output*. These values range from a minimum of 1,500 drams (15.63 lbs.) to a maximum of 24,000 drams (250 lbs.). Harvey mentions only two of these values: 1,000 half ounces or 4,000 drams (41.67 lbs.), and 3,000 drams (31.25 lbs.). However, he does emphasize the most important fact about all forty-eight values, namely, the calculated half-hour cardiac output* is "*a larger quantity in every case than is contained in the whole body!*" (ibid.; emphasis added). Today, the average adult human body is thought to contain about 5.7 liters or 16 pounds (Apothecary) of blood (Considine 1983, 382). In the seventeenth century people were significantly smaller than today, so 14 pounds (Apothecary) would seem to be a plausible estimate of the actual seventeenth-century average.

Referring to table 2.1, one may wonder which of these forty-eight

TABLE 2.1 Harvey's Human Blood Flow Calculations (in Apothecary Units)

Ventricle Capacity	Eject Fraction	Pulse Rate	Stroke Weight	One-half Hour Cardiac Output*
2 oz.	1/4	33/min.	4.00 dr.	4,000 dr. (41.67 lbs.)
2 oz.	1/5	33/min.	3.20 dr.	3,200 dr. (33.33 lbs.)
2 oz.	1/6	33/min.	2.67 dr.	2,670 dr. (27.81 lbs.)
2 oz.	1/8	33/min.	2.00 dr.	2,000 dr. (20.83 lbs.)
2 oz.	1/4	66/min.	4.00 dr.	8,000 dr. (83.33 lbs.)
2 oz.	1/5	66/min.	3.20 dr.	6,400 dr. (66.67 lbs.)
2 oz.	1/6	66/min.	2.67 dr.	5,340 dr. (55.63 lbs.)
2 oz.	1/8	66/min.	2.00 dr.	4,000 dr. (41.67 lbs.)
2 oz.	1/4	100/min.	4.00 dr.	12,000 dr. (125.00 lbs.)
2 oz.	1/5	100/min.	3.20 dr.	9,600 dr. (100.00 lbs.)
2 oz.	1/6	100/min.	2.67 dr.	8,010 dr. (83.34 lbs.)
2 oz.	1/8	100/min.	2.00 dr.	6,000 dr. (62.50 lbs.)
2 oz.	1/4	133/min.	4.00 dr.	16,000 dr. (166.67 lbs.)
2 oz.	1/5	133/min.	3.20 dr.	12,800 dr. (133.33 lbs.)
2 oz.	1/6	133/min.	2.67 dr.	10,680 dr. (111.25 lbs.)
2 oz.	1/8	133/min.	2.00 dr.	8,000 dr. (83.33 lbs.)
3 oz.	1/4	33/min.	6.00 dr.	6,000 dr. (62.50 lbs.)
3 oz.	1/5	33/min.	4.80 dr.	4,800 dr. (50.00 lbs.)
3 oz.	1/6	33/min.	4.00 dr.	4,000 dr. (41.67 lbs.)
3 oz.	1/8	33/min.	3.00 dr.	3,000 dr. (31.25 lbs.)
3 oz.	1/4	66/min.	6.00 dr.	12,000 dr. (125.00 lbs.)
3 oz.	1/5	66/min.	4.80 dr.	9,600 dr. (100.00 lbs.)
3 oz.	1/6	66/min.	4.00 dr.	8,000 dr. (83.33 lbs.)
3 oz.	1/8	66/min.	3.00 dr.	6,000 dr. (62.50 lbs.)
3 oz.	1/4	100/min.	6.00 dr.	18,000 dr. (187.5 lbs.)
3 oz.	1/5	100/min.	4.80 dr.	14,400 dr. (150.00 lbs.)
3 oz.	1/6	100/min.	4.00 dr.	12,000 dr. (125.00 lbs.)
3 oz.	1/8	100/min.	3.00 dr.	9,000 dr. (93.75 lbs.)
3 oz.	1/4	133/min.	6.00 dr.	24,000 dr. (250.00 lbs.)
3 oz.	1/5	133/min.	4.80 dr.	19,200 dr. (200.00 lbs.)
3 oz.	1/6	133/min.	4.00 dr.	16,000 dr. (166.67 lbs.)
3 oz.	1/8	133/min.	3.00 dr.	12,000 dr. (125.00 lbs.)
1.5 oz.	1/4	33/min.	3.00 dr.	3,000 dr. (31.25 lbs.)
1.5 oz.	1/5	33/min.	2.40 dr.	2,400 dr. (25.00 lbs.)
1.5 oz.	1/6	33/min.	2.00 dr.	2,000 dr. (20.83 lbs.)
1.5 oz.	1/8	33/min.	1.50 dr.	1,500 dr. (15.63 lbs.)

TABLE 2.1 *(continued)*

Ventricle Capacity	Eject Fraction	Pulse Rate	Stroke Weight	One-half Hour Cardiac Output*
1.5 oz.	1/4	66/min.	3.00 dr.	6,000 dr. (62.50 lbs.)
1.5 oz.	1/5	66/min.	2.40 dr.	4,800 dr. (50.00 lbs.)
1.5 oz.	1/6	66/min.	2.00 dr.	4,000 dr. (41.67 lbs.)
1.5 oz.	1/8	66/min.	1.50 dr.	3,000 dr. (31.25 lbs.)
1.5 oz.	1/4	100/min.	3.00 dr.	9,000 dr. (93.75 lbs.)
1.5 oz.	1/5	100/min.	2.40 dr.	7,200 dr. (75.00 lbs.)
1.5 oz.	1/6	100/min.	2.00 dr.	6,000 dr. (62.50 lbs.)
1.5 oz.	1/8	100/min.	1.50 dr.	4,500 dr. (46.88 lbs.)
1.5 oz.	1/4	133/min.	3.00 dr.	12,000 dr. (125.00 lbs.)
1.5 oz.	1/5	133/min.	2.40 dr.	9,600 dr. (100.00 lbs.)
1.5 oz.	1/6	133/min.	2.00 dr.	8,000 dr. (83.33 lbs.)
1.5 oz.	1/8	133/min.	1.50 dr.	6,000 dr. (62.50 lbs.)

calculations led Harvey to his 4,000 drams (1,000 half ounces) and 3,000 drams values for half-hour cardiac output*. Only seven of the forty-eight calculations yield these values. Since Harvey expresses it as 1,000 half ounces, the larger figure undoubtedly results from an underlying pulse rate of 33 beats per minute, that is, 1,000 beats per half hour. Thus we can trim the computational routes to the larger figure down to two: 2-ounce ventricle capacity with eject fraction 1/4, or 3-ounce ventricle capacity with eject fraction 1/6. Since the value of 2 ounces comes first in Harvey's enumeration of hypothetical ventricle capacities just as the value 1/4 comes first in his list of possible eject fractions, it seems virtually certain that Harvey obtained his 4,000 drams figure on the basis of a supposed 2-ounce ventricle capacity, a 33 beats per minute pulse rate, and an eject fraction of 1/4.

Similarly, the lower figure of 3,000 drams could have been derived in any of three ways, only one of which seems to be ruled out by the text. Harvey expresses this lower figure as "one thousand times three drachms", so we can assume an underlying pulse rate of 33 beats per minute (1,000 beats per half hour) and a stroke weight of 3 drams. But the assumption of a 3-ounce ventricle capacity with an eject fraction of 1/8 and the assumption of a 1.5-ounce ventricle capacity with an eject fraction of 1/4 both fulfill these conditions. Thus, we can be confident that Harvey calculated his lower value of half-hour cardiac output* in one of these two ways, but we do not know which.

In addition to the foregoing puzzle, Harvey's calculations also leave us with a *mystery,* but it is not the one signaled by Kilgour (quoted earlier). The mystery is from where the value of 1 dram as a *calculated* stroke weight—the product of hypothetical ventricle capacity with hypothetical eject fraction—comes. A glance at table 2.1 shows that none of the forty-eight calculations contains such an entry. We suggest that either (1) Harvey made a simple computational mistake—common when the Apothecary System is used—obtaining 1 dram when he should have obtained 1.5 drams for stroke weight; (2) the typesetter misread Harvey's nearly illegible script; or (3) the text of *De Motu Cordis* has been corrupted. The Latin text (Harvey 1628) agrees with the English translation (Harvey 1952a) in the matter of the 1 dram value. But how could this anomalous value have gone so long unnoticed in such a classic work? Probably because Harvey *was* a klutz when communicating the results of his calculations!

To see that the 1 dram value for stroke weight must be anomalous or errant, consider the following. When combining the 1 dram value with a 33 beats per minute pulse rate, one gets 1,000 drams (10.42 lbs.) for the half-hour cardiac output*, a quantity of blood substantially less than the average amount of blood in the entire adult human body—which even in the seventeenth century was about 14 pounds on our estimate of seventeenth-century body size—contra Harvey's emphatic claim that all the calculations based on his hypothetical values lead to half-hour cardiac outputs* that are greater than this total amount of blood. So, we have a second and independent piece of evidence that either Harvey made a simple computational mistake or that the text of *De Motu Cordis* is corrupted.

Not long after enumerating the various assumptions ingredient in his forty-eight calculations, Harvey lists anew in chapter 9 some possible values for the eject fraction, namely, 1/3, 1/6, or 1/8. Only the fraction 1/3 is novel, but it too fails to explain the anomalous entry of 1 dram for stroke weight, for 1/3 of the assumed left-ventricle capacities of 2 ounces, 3 ounces, and 1.5 ounces gives stroke weights of 5.33 drams, 8 drams, and 4 drams, respectively.

Ever the comparative anatomist, Harvey bids us consider the half-hour cardiac output* in dogs and sheep under the assumption that only one scruple of blood is ejected into the animal's aorta during systole. Using a 33 beats per minute pulse rate, Harvey obtains 1,000 scruples (3.47 lbs.) of blood or, as he says, "about three pounds and a half of blood injected into the aorta; but the body of neither animal contains above four pounds of blood, a fact which I have myself ascertained in the case of the sheep" (1952a, chap. 9, 287). That is, even under Harvey's extremely conservative assumptions about stroke weight and pulse rate, nearly as much blood is pumped through the sheep's or dog's heart in one-half hour as is contained in its entire body.

But Harvey's rhetorical strategy—to show that even extremely conservative estimates of pulse rate and stroke volume lead to quantities of blood flow so large as to overwhelm the Galenic theory—is perhaps most evident in the following, "But let it be said that this [the quantity of blood, calculated to be pumped into the arteries by the left ventricle in half an hour] does not take place in half an hour, but in an hour, or even in a day; any way it is still manifest that more blood passes through the heart in consequence of its action, than can either be supplied by the whole of the ingesta, or that can be contained in the veins at the same moment" (ibid.). In other words, Harvey claims—correctly—that even if he had used values of stroke weight and pulse rate that generate an absurdly small value for human cardiac output*— *about fifteen hundred times smaller than the average value accepted today*—he would still have made an irrefutable case against the Galenic theory. Even such a relatively tiny amount of blood flow was too much for the Galenic theory to handle. That he chose even to mention explicitly a value of half-hour cardiac output* only about 1/36th less than the accepted value today shows that he did not want to make his conservative assumptions so minuscule as to risk their being scorned as ludicrous. (Note that the smallest value of half-hour human cardiac output* generated by Harvey's forty-eight hypothetical calculations— 1,500 drams—is about 1/24 of today's lowest accepted value.)

After presenting the assumptions that generate the forty-eight calculations, Harvey gives a few more illustrative calculations of half-hour cardiac output* for humans, oxen, and sheep, again using a pulse rate of 33 beats per minute (ibid.; see table 2.2). The same anti-Galenic moral is implicit in these extremely conservative calculations. That Harvey nowise intends either the assumed values of pulse rate and stroke weight or the cardiac output* calculated therefrom to be realis-

TABLE 2.2 Cardiac Output* in Humans, Oxen, and
Sheep (in Apothecary Units)

Pulse Rate	Stroke Weight	One-half Hour Cardiac Output*
33 beats/min.	1 dr.	1,000 dr. (10.42 lbs.)
33 beats/min.	2 dr.	2,000 dr. (20.83 lbs.)
33 beats/min.	4 dr.	4,000 dr. (41.67 lbs.)
33 beats/min.	8 dr.	8,000 dr. (83.33 lbs.)

tic or accurate is patent from the passage that immediately follows his presentations of these calculations, "The actual quantity of blood expelled at each stroke of the heart, and the circumstances under which it is either greater or less than ordinary, I leave for particular determination afterwards, from numerous observations which I have made on the subject" (ibid.).

That Harvey is fully aware of the variability of human and animal pulse rates and cardiac outputs*, and of many of the factors on which they depend, is evident from the following passage in which he recapitulates his case against the Galenic theory:

Meantime this much I know, and would here proclaim to all that the blood is transfused at one time in larger, at another in smaller quantity; and that the circuit of the blood is accomplished now more rapidly, now more slowly, according to the temperament, age, &c. of the individual, to external and internal circumstances, to naturals and non-naturals—sleep, rest, food, exercise, affections of the mind, and the like. But indeed, *supposing even the smallest quantity of blood to be passed through the heart and the lungs with each pulsation, a vastly greater amount would still be thrown into the arteries and whole body than could by any possibility be supplied by the food consumed; in short it could be furnished in no other way than by making a circuit and returning.* (Ibid.; emphasis added)

In chapter 10 of *De Motu Cordis,* Harvey surveys some possible objections to his theory. Of these the most serious is the observation that a productive milk cow can produce seven or more gallons of milk per day from ingested food. But then why can our bodies not produce from the food we eat all the blood demanded by Harvey's observations? Harvey is quick to respond that "the heart by computation does as much and more in the course of an hour or two" (ibid., chap. 10, 288). But just which calculations lie behind Harvey's reply? Note that a stroke weight of 4 drams (1/2 ounce) and a 66 beats per minute pulse rate yield about 8 gallons of blood in one hour. The same stroke weight coupled to a 33 beats per minute pulse rate gives the same quantity of blood in two hours. It is probable, therefore, that these are the very calculations that led Harvey to claim that the heart does *in an hour or two* what the productive milk cow does in a day. Consequently, if the Galenic theory were correct, a human being would have to consume ten to twenty times more food than does a productive milk cow in order to produce the quantity of blood that conservative assumptions about pulse rate and stroke weight show must flow through the heart. Once again, the Galenic theory is swamped by numbers generated from extremely conservative hypotheses.

Even Kilgour recognizes that Harvey succeeded in making his case, his allegedly faulty value-assumptions notwithstanding. Harvey would not have succeeded equally well had he used more realistic assumptions about pulse rate and stroke volume in his computations because opponents would have focused on these assumptions rather than on his circulation theory, mocking them as incredible. For example, realistic assumptions would entail that the blood in the human body makes a complete circuit in about a minute or two—today we know that the entire supply of blood in an average adult human being at rest circulates in about one minute (Considine 1983, 1462), a fact that educated people find difficult to accept even now. Only after one has been habituated to Harvey's circulation theory can one begin to take in stride the large values of various associated phenomena, for example, the fact that during an average human lifetime the heart pumps almost enough blood to fill a large football stadium (ibid.). Furthermore, realistic assumptions might have seemed to imply that, once an artery or vein has been opened, all the blood would drain from the human body in a minute or so, a span much smaller than Galen's half hour.

By employing conservative values, Harvey diverted attention from the occasional phenomenon he had trouble explaining to the many phenomena he could explain easily and elegantly. For example, Harvey exposes as an artifact of the laboratory the fact that "in our dissections we usually find so large a quantity of blood in the veins, so little in the arteries; wherefore there is much in the right ventricle, little in the left" (1952a, chap. 9, 288), a fact that probably caused "the ancients to believe that the arteries (as their name implies) contained nothing but spirits during the life of an animal" (ibid.). Harvey's sophisticated explanation of the aforementioned phenomenon goes thus:

[As] there is no passage to the arteries, save through the lungs and heart, when an animal has ceased to breathe and the lungs to move, the blood in the pulmonary artery is prevented from passing into the pulmonary veins, and from thence into the left ventricle of the heart. . . . But the heart not ceasing to act at the same precise moment as the lungs, but surviving them and continuing to pulsate for a time, the left ventricle and arteries go on distributing their blood to the body at large and sending it into the veins; receiving none from the lungs, however, they are soon exhausted, and left, as it were, empty. But even this fact confirms our views, in no trifling manner, seeing that it can be ascribed to no other than the cause we have just assumed. (Ibid.)

Similarly, Harvey notes that his theory readily explains that "after death, when the heart has ceased to beat, it is impossible by dividing either the jugular or femoral veins and arteries, by any effort to force out more than one half of the whole mass of the blood. Neither could the butcher, did he neglect to cut the throat of the ox, which he has knocked on the head and stunned, until the heart had ceased beating, ever bleed the carcass effectually" (ibid.).

A Defense of Harvey Against Other Charges

One remaining task is to answer Kilgour's charge that Harvey mounted a bumbling defense of his second thesis that the blood flowing through the arteries into the extremities far exceeds both what is needed for nutrition and what the whole mass of body fluids could supply. Kilgour complains that Harvey "used no specific measurements and argued largely by inference" (1987, 14). True enough, but Harvey had no more need to use "specific measurements" here than he had to use realistic assumptions to establish his first thesis. All that Harvey needed to do was to invoke again his calculations of hypothetical cardiac output*, for, "upon grounds of calculation, with the same elements as before, it will be obvious that the quantity can neither be accounted for by the ingesta, nor yet be held necessary to nutrition" (Harvey 1952a, chap. 11, 289).

It was pointed out in the preceding section that the 1 dram value is errant. None of Harvey's forty-eight calculations yields a stroke weight of 1 dram; the closest value they generate is 1.5 drams. But Kilgour's attribution to Harvey of an estimate of 3.9 grams for human stroke weight is predicated on this 1 dram value (1 dr. = 3.888 g.). How, then, does Kilgour derive this figure?

First and most probably, Kilgour may have taken Harvey's mention of a calculated stroke weight of 1 dram at more than face value. Thereupon, in his eagerness to convict Harvey of blunders, Kilgour misconstrued this conservative hypothetical value as Harvey's own estimate of actual stroke weight (in the average adult human at rest). But it is, of course, no such thing. The 1.5 dram value—the closest calculated value to the errant 1 dram—represents the mathematical consequence of highly conservative assumptions about ventricle capacity and eject fraction. Thus, it in no way represents Harvey's own view about average stroke weight in adult humans.

Another possible source of Kilgour's 3.9 gram stroke-weight charge against Harvey is Harvey's comparative analysis of the cardiac outputs* of humans, oxen, and sheep under various assumptions about pulse rate and stroke weight, the results of which are displayed in table 2.2.

The fact that it happens to be the smallest of the four hypothetical values of stroke weight is no reason to take 1 dram to be Harvey's own estimate of stroke weight in the average resting adult human, ox, or sheep. Moreover, these cross-species calculations of half-hour cardiac output* occur in a context where Harvey is manifestly positing some exceptionally conservative values of stroke weight. Nor from the fact that Harvey here uses only one value for pulse rate for all three species—33 beats per minute—should one take this figure to be his estimate of average pulse rate in healthy adults of all three species. Like the small values for stroke weight, this pulse rate is meant to be conservative, and for exactly the same rhetorical-scientific reasons we encountered in the preceding section.

Note in table 2.3 that unlike dogs, goats, and horses—animals with which Harvey is familiar and which he mentions now and then in *De Motu Cordis*—oxen and sheep exhibit approximately the same pulse rates as humans (animal values are taken from Fraser et al. 1991, 966).

If Harvey could not even take an accurate pulse, is it not remarkable that he just happened to lump humans with those animals which have similar pulse rates? We can now also see why Harvey used a pulse rate of 33 beats per minute in his blood flow calculations about humans, oxen, and sheep. For all three of these species, *33 beats per minute is the best possible lower bound for pulse rate.* It is the only lower bound

TABLE 2.3 Pulse Rates and Ranges for Selected
Species

Species	Average Pulse Rate	Pulse-Rate Range (Healthy Specimens)
Dog	NA	100–130
Goat	90	70–135
Horse	44	23–70
Humans	72	60–85
Oxen	NA	60–70
Sheep	NA	60–120

that will generate an integral multiple of one thousand (when rounded off) for the number of heartbeats in one-half hour, thereby simplifying computation of half-hour cardiac output*.

Kilgour's Recantation?

So far we have debunked the derogatory claims and specious arguments of Kilgour's (1987) influential essay, originally published in 1952. But in 1954, Kilgour published another influential article on Harvey's use of the quantitative method (see Kilgour 1981), which might appear to be a recantation of some of his more defamatory, earlier charges against Harvey.

Kilgour (1981) does indeed incorporate some advances over his earlier article. First, the latter article has its medical terminology mostly straightened out. It employs the term cardiac output in almost its correct sense and happily begins to use the concept of stroke volume, although in both cases the latter article takes the respective units to involve blood weight instead of blood volume, a peccadillo compared to the terminological transgressions of his earlier piece.

Although Kilgour (1981) does not recant his earlier charges that Harvey was an experimental klutz and an indifferent advocate of his own theses, Kilgour does not repeat or renew these charges, at least not explicitly. Kilgour now concedes that the "reason why Harvey was not more precise in his measurements was that he did not have to be to demonstrate his great discovery" (ibid., 419), but the klutziness charge is nevertheless implicit in his remark that "[if] Harvey had been trained to use instruments of precision to control quantitative methods, he probably would have used an accurate pulse rate" (ibid.). Nor is the aforementioned concession altogether new. In his former article, Kilgour acknowledged that Harvey succeeded in making his case even with his "faulty" values. What is new in Kilgour (1981) is the realization, however dim, that Harvey was aware that he did not need realistic or accurate values to make his case. What is still missing is any realization on Kilgour's part that Harvey was acutely aware that highly conservative assumptions about pulse rate and stroke weight served his purposes in ways that more accurate or realistic ones could not have done.

Kilgour's earlier charge of experimental klutziness is supplanted by one of quantitative indifference. From the fact that Harvey knew that he did not need accurate values of pulse rate or stroke weight, Kilgour (1981) concludes that Harvey was careless or indifferent in his measurements of even something as simple as pulse rate:

Although he [Harvey] mentions pulse rates of 33, 67, 100, and 133 per minute, he used a rate of 33 in the only calculation in which he specifically mentioned man. It is obvious from this that he did not use his minute watch to make this measurement, and it is apparent from his lumping together "man, sheep or ox," all with a pulse of 33, which is reasonably correct for an ox but only half the correct figure for men or sheep, either that he did not take the trouble to measure their pulse rates, or that he was not particularly concerned about using the correct rate. Of the two alternatives, the latter is the more likely. (P. 416)

From our earlier discussion, Harvey's real reasons have about as little to do with carelessness and indifference as one could possibly imagine. As F. R. Jevons (1981) exclaims, "[N]ot giving specific quantities, or even using quantities not measured but only assumed to be reasonable, is not the same as being content with rough values because they are adequate to prove the point" (p. 465).

Kilgour now labels Harvey's "estimates of the stroke volume [stroke weight] as being either 1/4, 1/5, 1/6, or 1/8 of the ventricular content . . . *sheer guesses*" (1981, 416; emphasis added). But far from being shear guesses, these values for the eject fraction (and ventricular capacity, and therefore cardiac output*) were chosen by Harvey expressly for their conservative character.

In stark contrast to his disparaging treatment of Harvey, Kilgour lauds Richard Lower (1631–1691) for using "the total weight of the blood in the left ventricle as being the amount ejected during systole, instead of some fraction thereof as Harvey had assumed" (ibid., 421).

Through his comparative anatomical work, Harvey undoubtedly knew that a left heart never ejects its full charge during systole, but that a significant reservoir of blood remains in the left heart throughout its action. This reservoir today is taken to be about 2/5th of left-ventricle capacity in animals, humans included (Fraser et al. 1991, 9), a value somewhat but not strikingly greater than the value 1/4 that Harvey seemed to favor as conservative but not too unrealistic. So rather than castigating Harvey for his eject fraction values, Kilgour should have praised Harvey for having detected the existence of a ventricular reservoir. And instead of commending Lower, Kilgour should have chastised him for having ignored or overlooked this same reservoir. Presumably, what moves Kilgour to praise Lower but criticize Harvey is the fact that Lower's 2-ounce (16 dr.) value for stroke weight happens to be closer to today's accepted value than even Harvey's most liberal assumption about stroke weight (6 dr.). But Kilgour seems oblivious to the fact that Lower's closer value comes about only *per accidens*.

Another advance over his earlier position is Kilgour's belated realization that Harvey's computed 1 dram value for stroke weight is anomalous:

[O]ne dram is not one of the possible values, yet Harvey used it in his first and third calculation. Either he did not do the arithmetic . . . or he merely rounded off one and a half drams to one dram. Moreover, he included sheep in his second and third calculations and employed variously a cardiac output [stroke weight] of one scruple [1.3g], one dram [3.9g], two drams [7.8g], half an ounce [15.6g], and one ounce [31.1g]. Certainly Harvey was not concerned with accurate measurement. (1981, 417)

When he suggests that Harvey may not have done the arithmetic, Kilgour perhaps means, not that Harvey neglected to do it, but that Harvey may not have done it *correctly*. However, Kilgour's other suggestion—that Harvey might have rounded off 1.5 drams to 1 dram— carries no plausibility. Harvey computes his various values of stroke weight (from assumed values of ventricular capacity and eject fraction) only to immediately multiply them by 1,000—as well as by 2,000, 3,000, and 4,000—to get half-hour cardiac output*. To round off in the way Kilgour suggests would have caused Harvey to obtain values of half-hour cardiac output* that range from 500 drams to 2,000 drams smaller than what the computed half-hour cardiac outputs* should be. Harvey does round off numbers when rounding them off is harmless and helpful, for example, by substituting 1,000 for 990 when converting a 33 beats per minute pulse rate into number of beats per half hour. But no evidence suggests that Harvey made the mathematical blunder of rounding off in the sort of context described previously.

Kilgour also touches upon the *milk-cow objection,* but only to distort this serious challenge to the circulation theory. Kilgour conjectures that Harvey responded to an objection to the effect that "the blood flowed out of the body in the form of other fluids such as milk from mammals" (ibid., 418). Harvey does discuss now and then the storage problem which his observations and calculations pose for the Galenic system, namely, where and how the body stores so much blood coursing through the heart into the arteries. But, *pace* Kilgour, the milk-cow objection is treated in chapter 10 of *De Motu Cordis* in the more familiar context of the production problem which his observations and calculations pose for the Galenic system, namely, how to account for the manufacture by the body of so much blood in so little time.

Astonishingly, J. H. Woodger thinks that Harvey does not sufficiently highlight the production problem:

It is curious that Harvey does not make his statement about the amount of fluid ingested, compared with the amount of blood passing through the heart, more emphatic. If we note that at the rate of 500 oz. of blood per half-hour we shall have 24,000 oz. in 24 hours, we see that almost eight times the weight of a man of 14 stone passes in 24 hours, and no man ingests more than eight times his own weight in 24 hours! Even this is an underestimate, because 1000 heart beats in half an hour is an underestimate. (1981, 92)

But formalistic trappings notwithstanding, Woodger hardly represents a close reading of *De Motu Cordis.* For example, Woodger fails to realize that Harvey uses the Apothecary System. Woodger's remark about "almost [i.e., not quite] eight times the weight of a man of 14 stone" would be correct if Harvey's units were those of the British System—24,000 ounces = 7.65 × 14 stone—but it is incorrect in the Apothecary System—24,00 ounces = 8.4 × 14 stone. Woodger also misinterprets the principal computational passage of chapter 9 to mean that left-ventricle capacity is *at least* 2 ounces, that the eject fraction is *at least* 1/4, and that the pulse rate is *at least* 33 beats per minute, so that the half-hour cardiac output* is *at least* 500 ounces (ibid., 82). Finally and dismayingly, Woodger dismisses Harvey's compelling comparative-anatomical arguments as "propaganda" (ibid., 92), a "use of persuasion rather than an appeal to reason" (ibid.; see also Ghiselin 1981, 314–15).

Kilgour further mishandles the milk-cow objection when he asserts that "Harvey answered this criticism by saying that by computation the heart pumps *in an hour or less* more than all the milk produced in a day, but he did not do any computation" (1981, 418; emphasis added). First, Harvey's text reads "in the course of an hour or two" (Harvey 1952a, chap. 10, 288), not "in an hour or less" as Kilgour paraphrases. (The Latin original runs thus: *Respondendum, quod cor tantundem, vel amplius, una hora, vel altera, computatione facta, remittere constet.* [1628, 47].) Second, as demonstrated earlier, we can reconstruct with considerable plausibility the actual calculations on which Harvey's "in the course of an hour or two" remark was probably based.

So, although neither the experimental-klutz charge nor the indifferent-advocate charge reappears *overtly* in Kilgour (1981), this later essay hardly amounts to a recantation by Kilgour of his earlier article. Terminological improvements aside, Kilgour (1981) is in many ways retrograde. At least his earlier article recognized that Harvey's calculations are somehow important to his overall case for the circulation of the blood. However, Kilgour's later article joins a growing chorus of commentators who downplay the quantitative elements in *De Motu Cordis* (see Kilgour 1981, 421; note that the text quoted supplies pagination of the original text, which is cited here). For example, F. R. Jevons recognizes that "Harvey's quantitative experiments were 'thought experiments' " (1981, 465) based on assumed values and observes that "Harvey . . . put much emphasis on his suggested calculations" (ibid., 467). Nevertheless, Jevons concludes that "it seems undeniable that the importance of his quantitative considerations, for his own work as well as for the history of biology in general, has been overemphasized" (ibid.).

Similarly, M. Ghiselin, while acknowledging that "Harvey is repeatedly praised for his quantitative experiments" (1981, 325), asks, "Why, if his experiments are quantitative, does only one observation [ventricle capacity]—one which can hardly be called an experiment—contain a measurement? Why does he give so little weight to this in his argument, and why is the measurement simply a crude estimate?" (ibid.). Ghiselin contends that Harvey's work is grounded in comparative anatomy much more than in calculation (ibid., 314–15) and disparages the computational arguments of *De Motu Cordis* as "a very early stage in the development of a quantitative approach" (ibid., 326). But Ghiselin unwittingly closes in on the mark when he observes that Harvey's "system is so constructed that measurement is not necessary to establish the thesis. ... Thus the merit of the work lies, not so much in using mathematics and measurement, but in getting along without them" (ibid.). Had Ghiselin said "precise" or "realistic measurement" and omitted the hyperbole about mathematics, he would have hit the bull's-eye.

G. Plochman (1981) clearly recognizes that the production problem typically forms the backdrop of Harvey's quantitative arguments, that is, that these arguments are directed "against the Galenists, who asserted that the blood was formed in the liver from ingested food and renewed at every meal" (p. 204). But unlike Kilgour and Woodger, Plochman hits the bull's-eye with his passing remark that "all that Harvey is really saying is that while his figure may not be accurate, at least the blood flow cannot go *below* a certain minimum, which in turn is considerably above a certain other figure, the maximum blood that could have been converted from the intake of food" (ibid., 194).

REFERENCES

Anonymous. 1952. "Biographical Note: William Harvey, 1578–1657." In Hutchins, pp. 263–64.

Aquinas, T. 1923. *Summa contra Gentiles*. Rome: Forzanius et socius.

———. 1927. *De Aeternitate Mundi contra Murmurantes*. In R. P. Mandonnet, ed., *Opuscula Omnia*, Paris: P. Lethielleux, pp. 22–27.

———. 1952. *Summa Theologiae*. Rome: Marietti.

Berkow, R. et al., eds. 1987. *The Merck Manual of Diagnosis and Therapy*. 15th ed., Rahway: Merck & Co.

Cohen, I. B., ed. 1981. *Studies on William Harvey*. New York: Arno Press.

Considine, D., ed. 1983. *Van Nostrand's Scientific Encyclopedia*. 6th ed. New York: Van Nostrand Reinhold Co.

Davidson, H. 1987. *Proofs for Eternity, Creation and the Existence of God in Medieval Islamic and Jewish Philosophy*. Oxford: Oxford University Press.

Elkana, Y., and J. Goodfield. 1968. "Harvey and the Problem of the 'Capillaries.'" *Isis* 59: 61–73.

Ent, G. 1952. Untitled dedicatory preface to W. Harvey's *Anatomical Exercises on the Generation of Animals*. In Hutchins, pp. 329–30.

Fraser, C. et al., eds. 1991. *The Merck Veterinary Manual.* 17th ed. Rahway: Merck & Co., Inc.

Galen 1952. *On the Natural Faculties.* In Hutchins, pp. 167–215.

Ghiselin, M. 1981. "William Harvey's Methodology in *De Motu Cordis* from the Standpoint of Comparative Anatomy." Reprinted in Cohen with original pagination, from *Bulletin of the History of Medicine,* vol. 49 (1967), pp. 314–27.

Harvey, W. 1628. *Exercitatio Anatomica de Motu Cordis et Sanguinis in Animalibus.* Frankfurt: Guilielmi Fitzeri.

———. 1952a. *An Anatomical Disquisition on the Motion of the Heart and Blood in Animals.* In Hutchins, pp. 267–304.

———. 1952b. *Anatomical Exercises on the Generation of Animals.* In Hutchins, pp. 329–496.

———. 1952c. *A Second Disquisition to John Riolan, in which Many Objections to the Circulation of the Blood are Refuted.* In Hutchins, pp. 313–28.

———. 1952d. *The First Anatomical Disquisition on the Circulation of the Blood, Addressed to John Riolan.* In Hutchins, pp. 305–12.

Hoenen, M. 1990. "The Literary Reception of Thomas Aquinas' View on the Provability of the Eternity of the World in de la Mare's *Correctorium* (1278–9) and the *Correctoria Corruptorii* (1279-ca 1286)." In Wissink, pp. 39–68.

Hutchins, R., ed. 1952a. *Gilbert: Galileo: Harvey,* Great Books of the Western World, vol. 28. Chicago: Encyclopedia Britannica, Inc.

———. 1952b. *Hippocrates: Galen,* Great Books of the Western World, vol. 10. Chicago: Encyclopedia Brittanica, Inc.

Jevons, F. R., 1981. "Harvey's Quantitative Method." Reprinted in Cohen, with original pagination, from *Bulletin of the History of Medicine,* vol. 36 (1962), pp. 462–67.

Kilgour, F. 1981. "William Harvey's Use of the Quantitative Method." Reprinted in Cohen, with original pagination, from *Yale Journal of Biology and Medicine,* vol. 25 (1954), pp. 410–21.

———. 1987. "William Harvey." In *Scientific Genius and Creativity,* Readings from *Scientific American.* New York: W. H. Freeman & Co., pp. 11–15.

Morris, W., ed. 1969. *The American Heritage Dictionary of the English Language.* Boston: Houghton Mifflin Co.

Page, J. et al. 1981. *Blood: The River of Life.* Washington: U.S. News Books.

Plochman, G. K. 1981. "William Harvey and His Methods." Reprinted in Cohen, with original pagination, from *Studies in the Renaissance,* vol. 10 (1963), pp. 191–210.

Van Veldhuijsen, P. 1990. "The Question on the Possibility of an Eternally Created World: Bonaventura and Thomas Aquinas." In Wissink, pp. 20–38.

Wissink, J., ed. 1990. *The Eternity of the World.* Leiden: E. J. Brill.

Woodger, J. H. 1981. "An Analysis of Harvey's *De Motu Cordis et Sanguinis.*" Reprinted in Cohen, with original pagination, from *Biology and Language* (1952), pp. 75–92.

3

The Cognitive Functions of Scientific Rhetoric

Philip Kitcher

Department of Philosophy, University of California at San Diego

> Shall that be shut to Man which to the beast
> Is open? or will God incense his ire
> For such a petty trespass, and not praise
> Rather your dauntless virtue, whom the pain
> Of death denounced, whatever thing Death be,
> Deterred not from achieving what might lead
> To happier life, knowledge of good and evil?
> Of good, how just? of evil, if what is evil
> Be real, why not known, since easier shunned?
> God, therefore, cannot hurt ye, and be just;
> Not just, not God; not feared then, nor obeyed:
> Your fear itself of death removes the fear.
> Why, then, was this forbid? Why but to awe?
> Why but to keep you low and ignorant,
> His worshippers? He knows that in the day
> Ye eat thereof your eyes, that seem so clear
> Yet are but dim, shall perfectly be then
> Opened and cleared, and ye shall be as Gods,
> Knowing both good and evil, as they know.
> —Milton, *Paradise Lost*

Eve is standing, fascinated, before the tree whose fruit she has been forbidden to eat. Seduced by the words of the serpent, she attends to some aspects of her complex predicament, overlooking others—just as, two hundred or so lines later, Adam too will focus on some features of his situation, and ignore others. The serpent's rhetoric colors Eve's way of thinking about her possible actions. She stretches out her hand, and the Fall ensues.

Science is not to be like that. From the dawn of modern science strenuous announcements have advocated that thinkers are not to be diverted from the conclusions that they ought to reach by the enchantments of language used to clothe a chain of reasoning. Hence the need for a restrained idiom that will "let the facts speak for themselves." Science is to be a rhetoric-free zone.

This once-orthodox conception of science and its language thrives on a contrast between surface and depths. We have the linguistic performances in which scientists engage; behind the performances is the reasoning that occurs in the scientists' minds. The danger of rhetoric is that the character of the surface may distort a scientist's vision of the depths.

A popular alternative conception challenges the existence of standards according to which underlying processes of reasoning can be assessed. There is no stripping away the surface to disclose the depths and to evaluate their epistemic worth. Instead a proper—naturalistic—study of modes of belief formation and belief change can only focus on the actual factors that produce doxastic commitments. Among these, the modes of presentation of ideas, the linguistic forms employed by scientists, play an important role, and it is important to study how they operate in science as elsewhere. Without raising the question of original sin, we can ask what rhetorical devices persuade members of various scientific communities.

Both conceptions contain insights and oversights. In what follows I articulate a view of scientific rhetoric that combines elements of each. I share with the second approach skepticism about the possibility of transparent ways of inducing belief, of devising a scientific idiom that is free of rhetoric. Modes of presentation have causal consequences. Nonetheless, I share with the first approach the view that one ought to reason in prescribed ways. Consequently, a serious question arises about whether the modes of presentation actually employed contribute to the proper formation of belief. Without invoking the notion of logical depths that can be distorted by rhetorical surfaces, I propose the possibility of evaluating the rhetoric used by scientists in terms of its tendency to promote good doxastic consequences. On this view scientific language works well not because it is logically transparent but because it encourages valuable epistemic performances.

Exemplifying a theme which I touch on later, I clarify my position by inventing a series of opponents of increasing sophistication, all of whom are impatient with my "obsession with normativity." The first adopts an extreme position: There are no constraints whatsoever on belief formation. Imagine a scientist, the *source*, who makes a linguistic presentation to another scientist, the *target* (see Kitcher 1990b for discussion about how scientists persuade themselves, i.e., source and target might be identical—a topic ignored for present purposes). According to my extreme opponent, no epistemic grounds exist for appraising the response that target makes to source's performance—we cannot describe the target as enlightened or dazzled, instructed or bewitched. Anything goes.

Now I suppose that source's performance causes target to undergo a psychological process, a sequence of representational states. This particular process is classified as a psychological type, as we abstract from some of the specific details of the representations induced.[1] The epistemic quality of target's response is to be understood in terms of the propensity of processes of that type to issue in beliefs of the kinds to which scientists aspire. Scientists exert considerable effort so that they and their colleagues will come to hold beliefs with certain desirable qualities. Because of the characteristics of human thought and cognition, and because of the way the world works, certain types of psychological processes are more likely than others to issue in the formation of beliefs with these desirable qualities. Because of the characteristics of human thought and cognition, certain types of linguistic performances by sources are more likely to lead targets to undergo processes that are more effective in fixing valuable beliefs. So it is not the case that anything goes.

My imaginary opponent becomes more sophisticated. We can concede the possibility of *local* standards that provide a basis for *local* criticisms. Within a particular community, certain types of beliefs are valued and we may investigate the kinds of processes that lead to the fixation of such beliefs and the uses of language that induce such processes. Such investigations depend on taking the community's values for granted, and presupposing the community's conception of the characteristics of human thought and of the world that human inquirers attempt to fathom. We have, however, no way of transcending particular communities and finding a standard to which *all* human thought and language can be held accountable. So, at least, claims my imaginary opponent.

Now of course the standards we actually employ in assessing the performances of ourselves and others depend on what we take to be valuable and what we believe about the world's workings. How could they not? However, in employing these standards, we recognize the possibility of our own error. Independently of us and our beliefs, certain types of doxastic achievements are valuable and certain features of the world obtain that are relevant to our pursuit of those doxastic achievements. Behind the local standards employed in particular communities stands the overarching standard of framing our thinking and our language so that they promote the attainment of the epistemically valuable, and we have an important and serious task of trying to correct the shortcomings of our local standards. This task becomes visible in the history of methodology—as, for example, when particular scientific groups demand that experiments for settling some issues be carried out according to a specified design. I claim that it can also be discerned in the history of the conventions of scientific dialogue that the forms of language prescribed or proscribed in scientific presentation are products of an attempt to satisfy the overarching standard.

At this point I anticipate a number of objections. One is to deny the relevance of the way the world is—an objection easily answered. If one wants to attain a goal, any goal, improvement of one's chances of doing so depends on conforming action to the way things are, not to the way they are believed to be. My point is simply the restriction of this to cases in which the goal is to achieve certain kinds of beliefs. A deeper objection would be to deny the possibility of ever knowing enough about the workings of the world to undertake the reform of our belief-forming processes. Yet this skeptical stance seems belied by the histories of methodology and of scientific rhetoric. Surely we do know enough about ourselves and the world to recognize that predictions formed on the basis of neglecting the base rate or committing the gambler's fallacy are likely to fail, and that sources can be gulled into overlooking relevant evidence by language that charms by its novelty or that tickles their vanity. Milton's portrait of the serpent's seduction owes its effect to knowledge of human proclivities that is readily obtainable by all of us in our mundane encounters with others.

The most serious opposition is articulated by conceding that, once a goal has been fixed, the ultimate standard for evaluating our efforts must take into account facts about the working of the world, but to deny a unique conception of epistemic value. This denial has recently taken two different forms. One version, championed by L. Laudan (1984), turns to the history of science to defend "axiological pluralism." According to Laudan, at different periods in the history of science, different scientific communities have valued different epistemic goals. In particular, Laudan is impressed with the debate between instrumentalists and realists about the characteristics of acceptable theories.

Whereas Laudan's discussions of the aims of science presuppose that one can separate epistemic values from other values (while allowing for a plurality of epistemic values), the other prominent way of attacking a unique conception of epistemic value seeks to undercut the distinction between the epistemic and the nonepistemic. S. Shapin and S. Schaffer (1985) develop a view that is often put forward far more vaguely when they argue for the inextricability of epistemic, moral, and social values in the debate between Boyle and Hobbes. This type of challenge can be understood as disallowing the possibility of delineating a field of activity within which science operates to aim at values that are distinctively its own, and, correlatively, that as many ways are possible to identify "epistemic" values as ways to identify "forms of life" which communities may adopt.

Both positions seem to ignore the possibility of analyses that will disclose the commonality of values underlying a more superficial diversity. Laudan's prime example, the debate between realists and instrumentalists, can best be understood by supposing that all parties share a conception of epistemic value: All would like to achieve true statements that would explain the phenomena. Their differences arise from divergent views about what is possible for scientists to achieve. Similarly, I suggest, the quarrel between Boyle and Hobbes masks a shared commitment to the epistemic goal of attaining accurate representations of nature and its structure, conceived both as valuable in themselves and as useful for the attainment of more practical ends.

These remarks only indicate the lines along which I would defend the thesis of a unique conception of epistemic value, at the core of which is the notion of explanatory truth (see Kitcher 1993). Underlying the various instrumental and derivative ends that scientists belonging to different communities may have is the goal of producing true statements that reveal the structure of nature. The episodes of thinking, speaking, and writing in which scientists engage can be evaluated in terms of their conduciveness to the elaboration of explanatory truths. So far I have presented this in individualistic terms, thinking about the ways in which the belief formation of a particular person, the target of someone's persuasive performance, should be directed in order to attain such epistemic goals. However, the achievement should more properly be attributed to the community, and our concern ought to be with the ways in which beliefs become broadly accepted among communities of inquirers. Methodology and rhetoric have the joint tasks of understanding how the numerous linguistic interactions among scientists and the processes of reasoning which they provoke should proceed if the group of scientists is to be successful in arriving at explanatory truths. The contrast between the individual and the social perspective can be best appreciated by noting that a rhetorical device might serve a cognitive function even if it prompted an individual member of the community to engage in methodological errors, provided that these errors promoted the communitywide quest for explanatory truth. I return to this possibility much later (the possibility is akin to points made in Kitcher 1990a about the mismatch between individual and community rationality).

So far I have been developing my position in dialogue with one of the main contemporary views about the role of rhetoric in science. Let me now turn to the other. Given that one believes that the function of rhetoric is to promote the search for explanatory truth, it is natural to suggest that the ideal form of scientific discourse should enable the logical structure of scientists' reasoning to become transparent to others, and that is done by eliminating all rhetorical devices in favor of a plain idiom. But I believe that the notion of a plain idiom in which logical structures are presented naked is as illusory as that of a language of pure observation in which the deliverances of the senses stand bare. I develop this point by turning to the area in which the construction of a logically transparent language seems to have been carried out most successfully.

Standard presentations of mathematical proofs are designed to conform to conventions familiar to professionals. They are broken down into lemmas, some conclusions are characterized as "obvious" or as obtainable by applying some specified strategy (taken to be well known to the reader), and, frequently, explicit commentary is offered about the general shape of the proof. From a logical point of view, the employment of these rhetorical techniques is unnecessary: The ancillary remarks mentioned are what G. H. Hardy (1940) described as "gas." Those who dismiss the idea of rhetoric within mathematics suppose that we could replace existing proofs without cognitive loss. They see the presentations actually given as enabling the professional mathematician to undergo the same process of reasoning as would be induced by reading a full proof in formal logic (or as enabling the mathematician to construct such a proof). The same admirable cognitive state could in principle be produced, at cost to the environment, by writing out rhetoric-free proofs in the language of formal logic. "Gas" is simply a cognitively effective way of saving the trees.

I contend that this orthodoxy about mathematical proofs is wrong. The formal proof does not serve the same cognitive function as the standard presentation it is supposed to replace. We have an optimistic vision of the powers of the cognitive subject, imagining that, once faced with a rigorous formal proof, the subject cannot fail to assent to any of the steps and thus can have no grounds for doubting the conclusion. By the same token, given a faulty formal proof, subjects will invariably spot the error and resist the conclusion. However, these comfortable assumptions about ourselves are belied by the pedagogy and the practice of mathematical logic. Logic teachers are keenly aware that students often misidentify the logical structures of relatively simple statements—by failing to recognize the pairing of parentheses, for example. These errors can be corrected by introducing a minimal rhetorical device (labeling parentheses), and, eventually, the student develops skill at identifying simple logical structures.

However, I suggest that trained students and professionals would face analogous problems when dealing with the kinds of formulas that would translate interesting mathematical theorems. Given our perceptual capacities, each of us would have trouble with a translation of, say, the mean-value theorem into the primitive language of first-order set theory. (To appreciate this point, try plunging into the middle of *Principia Mathematica* (1925), which, of course, employs abbreviatory conventions.) Moreover, we have a further difficulty. Lacking any explicit commentary, formal proofs typically leave one unable to understand how the proof works. They thus fail to discharge the explanatory functions fulfilled by standard presentations, and, I claim, this explanatory failure infects their ability to induce certainty. For, faced with a lengthy derivation of a surprising theorem, full of complex logical formulas, a responsible cognitive subject should recognize her fallibility in perceiving the logical structures of the formulas, and having no explanatory text to dispel the original sense of surprise, should wonder if this is not one of those occasions on which her capacities for identifying logical relations have failed her.

I maintain that the formal proof and the ordinary presentation have complementary functions. Standard mathematical proofs clear up many doubts about the truth of the theorem and provide insight as to why the theorem is true. *Once we have the standard proof,* a formal reconstruction of it can remove residual doubt that there is a lurking logical lacuna. Its power to do so is, I suggest, parasitic on the informal proof with its supposedly redundant commentary. So, if I am correct, even in mathematics, the imperfections of the cognitive subject make it impossible to serve all the cognitive functions of proof by constructing a rhetoric-free language. There is a serious subject of understanding how various ways of presenting mathematical proofs contribute to or detract from the cognitive achievements of those who study such proofs. "Mathematical rhetoric" is no oxymoron.

When we turn from mathematics to sciences such as physics, chemistry and biology, no well-established language represents proper scientific reasoning, and thus there is no possibility of demonstrating that the candidate logically transparent language fails to serve some cognitive functions. Nonetheless, I argue that the imperfections of the cognitive subject provide an even greater field for rhetorical devices, and I will try to understand some of the conventions of public scientific discourse as responses to recognitions of human cognitive shortcomings.

In principle, the study of rhetoric in science ought to embrace both public texts and the myriad conversations and exchanges that stand behind them. For present purposes I identify only cognitive roles that rhetorical devices play in published scientific documents, and I focus on two types of texts as prototypes of a range of printed materials. The first of these is the article that reports a new experimental finding. The second is the article or book that proposes a large new theoretical perspective. For exemplars of the first we can review issues of *Nature, Science,* or any of a host of more specialized journals. Lavoisier's memoirs on combustion and calcination or Darwin's *Origin of Species* ([1859] 1964) serve as representatives of the second.

Readers of experimental reports face two types of cognitive tasks. First, and more obviously, they have to accept or doubt the alleged finding—ideally being convinced when the experiment has been well done, and remaining dubious when it has not. The second task is to see how the reported result is relevant to the findings of others and to the reader's own work. Let us see how the current conventions of scientific presentation contribute to the performance of these tasks.

In assessing the status of an alleged experimental finding, readers must eliminate two sources of trouble, incompetence and fraud. Until recently, the latter problem has received scant attention, either because it has been supposed that the high calling of science attracts people who are unusually honest or because the mechanism of replication is assumed to be an efficient detector of fraudulent data. Recent exposure of instances of fraud in a variety of areas of experimental research has cast doubt on both assumptions. (For discussions and responses, see Merton 1973, Collins 1985, and Engler 1987.)

How does a reader judge that an experiment has been thoroughly performed? Part of the answer lies in the system of scientific authority. Those who submit articles to professional journals typically have institutional affiliations which are obtainable only by undergoing extensive programs of training. Once an article is published, it inherits some of the authority of the journal and of editors and referees. However, part of the answer also lies in features internal to the article itself: in the author's ability to display experimental virtuosity to those who undertake similar experiments. It is now a thoroughly familiar point that descriptions that would convey to a wide audience enough information to enable members of that audience to perform an experiment for themselves would be enormously long. Even specialists find that they cannot perform some experiments reported by others without recourse to telephone calls or face-to-face interactions. Nevertheless, reports of new experimental findings are required to contain extensive "Methods" sections. Such sections are supposed to convince interested readers that the author has the competence to perform the type of experiment that serves as the basis for the new finding.

Two main rhetorical devices serve in the establishment of proficiency. One is to anticipate and respond to worries about potential disturbing factors, to describe the ways in which the apparatus was shielded or one's reactants purified, for example. Another is to describe one's experiment in terms of established techniques, explaining, perhaps, how these were modified for the purposes at hand. The refereeing process is, in effect, a dialogue between author and referee, in which the latter tries to guide the use of these two strategies so as to increase the chances that members of the concerned community will accept the author's proficiency.

These conventions are well established within the scientific community, and the kind of rhetoric that they demand is naturally characterized as "dead rhetoric" by analogy with dead metaphors. Like clichés, the everyday rhetoric of scientists has become invisible. However, in some areas of experimental research, existing conventions of experimental reporting are being considered insufficient. A rash of well-publicized instances in which experimenters have misrepresented their findings has given rise to new debates about the establishment and maintenance of scientific credibility. Journals are never likely to publish pictures of animals "sacrificed"—or, even better, videotapes of the "sacrifices"—but they may demand that their authors allay suspicions that the data have been misrepresented. Some members of the biomedical community believe they face a problem about credibility—a problem that has a rhetorical dimension. What conventions should we adopt that would enable referees and editors—or at worst a more discerning readership—to better detect cases of experimental fraud?

I now address the second function that the language of a scientific article must discharge. I have spoken casually of the community to which an experimental report is addressed. But how do members of this community identify this report as speaking to them? The writer's task is to present the finding in a way that will quickly and clearly convey information about the kinds of scientific problems on which it bears.

I suppose that each scientist can generate a maplike representation that depicts the relations of the problems in the discipline in which she works. Without this ability the papers in most scientific journals would be as devoid of significance as they are for the outsider, who reasonably wonders why anyone should care about "evidence from stellar abundances for a large age difference between two globular clusters" (Nature 1991, 212–14) or "impairment of mitochondrial transcription termination by a point mutation associated with the MELAS subgroup of mitochondrial encephalomyopathies" (ibid., 236–39). One of the most prominent features of contemporary science is its division of domains into problems of extraordinary specificity, but, for the expert, the hierarchical structure of these problems can readily be appreciated and their relations to larger questions identified. So, for example, an apparently insignificant point about the possibility

of separating certain subcellular structures in mice may be seen as crucial for the project of identifying the time of expression of a particular
gene. That project, in turn, may be conceived as the best way of answering a question about the development of a particular neuronal
structure, and, ultimately, because mice are seen as an ideal system for
investigating issues in neuroembryology, this may bear on the largest
questions about the development of mammalian brains.

In general, scientists recognize that solutions to some problems are
needed to undertake others and they are able to relate specialized questions to issues of large cognitive or practical import. Moreover, they
can locate themselves and their own research on this map of the field—
typically, I think, in something like the way the famous *New Yorker*
Manhattan perspective represents the U.S.A. (This effect would be
similar to the kinds of salience effect described by Tversky, Kahneman,
Nesbet, Ross, and others. See Solomon 1992.) Different scientists are
likely to disagree in their perspectives on the field: Although, at some
level of generality, they share common problems, they favor different
routes to the solution. You may think that a sequence of studies on the
leech will resolve a large issue in neuroembryology, while I am committed to pursuing a series of inquiries on the mouse. When I announce
my breakthrough finding about expression of particular bits of mouse
DNA, you ought to pay attention—for people working along my line
of inquiry may now be markedly closer than you and your friends to
the solution we both seek, or, perhaps, my line of solution already
forecloses options that you would otherwise have thought worth pursuing. Hence my report should enable you to identify the common
ground represented in your map of the field and in mine, showing how
we have rival perspectives on this terrain.

An important function of the experimental report is to advertise
findings to those whose research is affected by them, both consumers
and devotés of rival lines of investigation. The initial paragraph will
often provide a context in which the experimental result is set, frequently alluding to a well-known piece of work that has partially resolved some problem of broad concern. By referring to particular figures and reviewing the recent history of inquiry, authors sketch the
geography of the subfield as they see it. (References do not just serve
as "allies" in Latour's 1987 sense, but as ways of focusing a picture of
the interrelations of problems. Also, as noted earlier, referring to others may play the role of demonstrating virtuosity.) Ideally, this rhetori-

cal work should be accomplished so that those whose research could benefit from the report will be drawn to it, while those to whom it is irrelevant will ignore it.

So far, the rhetorical devices on which I have focused are unexciting. Conventionalized forms of scientific discourse—dead rhetoric—provide little opportunity for creativity. However, on some occasions the rhetorical tasks facing scientists are far more challenging. When a scientist hopes to offer a large new theoretical perspective, it is far more difficult for potential readers to undergo the processes of reasoning that underlie the originator's espousal of the new view, and the fate of the novelty may depend on an author's ability to break down deep-seated habits of thought.

Consider the kinds of shift in scientific thought that we associate with Copernicus, Lavoisier, Darwin, Einstein, and Wegener. That the proposals of these thinkers faced large numbers of unsolved problems is a commonplace. In each case, it was necessary to forestall the objection that established explanations would have to be abandoned without any means of replacement. The new proposals are best understood as delineating strategies for solving problems, identifying the forms of reasoning that would enable scientists to answer the main questions of the field. Thus, in the case of Darwin, the *Origin* advances what I call "problem-solving strategies" that promise to explain phenomena of biogeography, comparative anatomy and physiology, the fossil record, and of comparative embryology, by constructing histories of descent with modification and to explain phenomena of adaptation—the fit of organism to environment—by detailing the operations of natural selection. For Darwin, as for the other figures, the task was to make vivid the promise of the chosen strategies in advance of working out the details.

This task can be broken up into subsidiary projects. First, the class of strategies proposed must be sufficiently comprehensive. To this end, the author needs to provide a conceptualization of the problems of the field that will encourage the conclusion that the strategies of solution can be implemented. Through two decades, culminating in his treatise of 1790, Lavoisier struggled to order the experimental results of chemistry so that they would be surveyable and so that the power of his problem-solving strategies would be apparent. Similarly, the *Origin* reconceptualizes problems of natural history, dividing them into recognizable types (biogeography, comparative morphology, and so forth) to which the Darwinian strategies have distinct applications.

Second, the new proposal must forestall apparent objections that would be likely to arise for those who consider implementing the proposed strategies. The difficulties that arise here stem from the nonmonotonic character of ampliative inferences. When the aspiring theorist offers an inductive argument, her audience may add an extra premise that saps its strength. When she claims that a phenomenon can be explained in a particular way, those critical of the claim may appeal to some background belief that is apparently inconsistent with the proposed line of explanation. When she hails her account as the only explanation, she invites the reader to search for rivals.

In all these instances, an ampliative argument is launched into a target's psychological life where it may quickly be amended by incorporating extra beliefs. The kind of reasoning in which the target engages will depend on the way in which the mode of presentation of the reasoning provokes or inhibits the addition of such beliefs. Use of a telling analogy, for example, may associate a potential defeating premise with the further information that would reveal that that premise is misleading.

These points can be recognized more precisely if we think of large novel scientific proposals as facing a pair of recurrent predicaments. *Inconsistency* predicaments occur when the proposal is apparently committed to a set of statements that are jointly inconsistent. *Explanatory-loss* predicaments arise wherever the proposal is unable to generate an instantiation of its problem-solving strategies to answer a question addressed by its competitor(s). These predicaments stand in an obvious relationship. Solving an inconsistency predicament by abandoning some statement in the inconsistent set typically leads to explanatory losses—one forfeits all the explanations in which the abandoned statement played a role. Conversely, efforts to make good explanatory losses often run aground on inconsistency predicaments.

Proponents of theoretical novelties face a host of such predicaments. To succeed they must show, at least in outline, how most, if not all, such predicaments can be resolved. This work is done by identifying lines of escape from the initial predicament, showing, for example, how abandoning a particular statement can resolve an inconsistency and how the losses incurred can be rectified in other ways. In general, we can think of inconsistency and explanatory-loss predicaments as associated with *escape trees*. These are branching structures that represent the possible ways of responding to the predicament. With respect to an inconsistent set of statements, for example, the initial branching is defined by the ways of deleting just one statement from the set (so if there are *n* statements in the set, there will be *n* initial branches). A branch is closed if it leads only to inconsistencies or to explanatory losses. To defend novel theoretical proposals one must show that the predicaments it faces have escape trees containing at least one open branch—or at least one branch that stands a good chance of being open.

Defense is only one part of the aspiring theorist's case. Equally important is identifying the shortcomings of rival points of view. Here too we may think in terms of inconsistency predicaments, explanatory-loss predicaments, and escape trees. Some of the most brilliant attacks in the history of science—for example, Galileo's critique of Aristotelian cosmology and physics—assemble one inconsistency predicament after another and show how all paths through the associated escape trees are closed. Beleaguered defenders are forced to confess, like Simplicio, that they must trust to some as yet unarticulated explanation.

Suppose, then, that we think of large scientific debates in terms of the recognition of inconsistency and explanatory-loss predicaments and the exploration of escape trees. From the methodological point of view, an ideal scenario would be if all parties to the dispute saw the structure of the situation, identifying all the predicaments for each of the rivals, all the closed paths, and all the open paths. This ideal is unattainable: Like Eve and Adam before us, we face situations whose complexity makes them vulnerable to the way in which language highlights particular features and casts others into shadow. What can be provided is a way of generating representations of the situation that will signal the presence of at least one open path from nodes that do not inevitably lead to inconsistency or explanatory loss and that will flag all paths from a node as closed when that is indeed the case. Furthermore, this way of generating representations of the cognitive options must deal with cases wholesale, finding general ways of classifying predicaments and general ways of indicating openness or closure for families of branches.

I can now finally make vivid the difference between my conception of rhetoric in science and the two alternatives with which I began. The traditionalist who worries that rhetoric will do to science what the serpent did to Eve is haunted by the vision of scientists who overlook or misidentify paths through escape trees because of the seductive language employed in advancing some scientific claim. The postmodernist, for whom rhetoric is everything, does not see the cognitive structure of scientific debates, missing the ways in which the systematic blocking off of options can force agreement. My pre–postmodern *via media* starts from the thought that the cognitive task facing those who would entertain a large new approach to some scientific domain is extraordinarily hard, so that the role of rhetoric is to guide them to an accurate representation of the structures of escape trees. Just as I claimed earlier that, without the kinds of information conveyed in informal proofs, strict formal derivations would be unintelligible, so too I now suggest that, without the creative use of rhetorical devices, scientists seeking to evaluate large theoretical novelties would be cognitively lost.

Having arrived at what I hope is a clear statement of my view, I now want to modify it. First, I believe that interactions among members of a scientific community frequently lead to improvements in the reasoning that is originally advanced to support a view. Darwin's exchanges with his critics led to the exploration of predicaments that Darwin had not originally considered and of lines of resolution that he had not previously seen. I claim that these interactions produced a body of texts that collectively induced reasoning in members of the community of British natural historians that was superior to the reasoning that had initially moved Darwin (and his original allies) to accept the evolutionary relatedness of organisms. Thus I suggest that rhetorical devices that misrepresent the cognitive structure of the situation may perform valuable cognitive functions if they lead to interactions among members of the community that ultimately disseminate superior reasoning.

Second, as M. Solomon (1994) has argued, a scientific community may modify its views because members of subcommunities respond to partial evidence. In the situation I have just envisaged, the community works together to craft a presentation of the case for a scientific view, and this presentation engenders a process of reasoning that rightly leads community members to adopt the view. But subcommunities might judge the merits of the view by focusing on a limited set of problems that are of local interest, and ultimately, all subcommunities might respond in this provincial way without anyone's ever formulating the global reasons for acceptance. Rhetorical presentations that illegitimately downplayed the significance of the problems that the view faced in some areas might be effective in attracting subcommunities not concerned with those areas so that the view might obtain representation within the broader community because it responded to the special interests of a subgroup. Moreover, this situation might even provide motivation for others to look more seriously at the prospects of the view in areas in which it faced outstanding problems, leading ultimately to successful resolution. Were this scenario to occur, then even rhetoric that *disguised* the cognitive structure of the debate at its initial stage might play a valuable role in the community's pursuit of truth.[2]

My discussion so far has been abstract, concerned with making room for possibilities that I think have been overlooked in recent discussions of rhetoric in science. I close with a brief (far too brief) look at two concrete examples in which creative scientific rhetoric serves a cognitive function.

Consider first Lavoisier's well-known campaign to displace the phlogiston theory and to establish his new chemistry. One crucial part of this campaign consisted in rendering the epistemic situation surveyable. There is something to the apparently peculiar suggestion that all the important arguments were waiting to be made long before anybody appreciated them. Consider the predicament of chemists after 1775. Most of the *cognoscenti* recognized at this stage that combustion involves the absorption of some constituent from the air. What remained at issue was whether there was also an emission of phlogiston. To resolve that question, it was necessary to explore the consequences of various proposals for a wide range of experiments, which *we*—after Lavoisier—would classify as combustions, reductions, acid-base reactions, and so forth. But too many results were available. Nobody could tell clearly if proposing an analysis which seemed to do nicely with some small family of reactions would wreak havoc when dealing with many others.

Lavoisier had difficulty in keeping the consequences of his own views straight. L. Holmes, in his exemplary discussion of Lavoisier's struggles to fashion a system of chemistry, notes that, in 1776, Lavoisier viewed himself as discovering something that he had apparently formulated one or two years earlier:

[Lavoisier's] apparent neglect of his own prior knowledge that respiration also produces fixed air is another manifestation of a mental characteristic we have repeatedly observed. When he concentrated his attention closely on a particular set of phenomena or relationships, he was apt to lose sight of other phenomena which, from a greater distance, appear obviously essential aspects of the problems he was attacking (1985, 66)

But this is no idiosyncrasy of Lavoisier's. Given the complexity of the material and the inadequacy of the terminology he was using to render it surveyable, Lavoisier simply had no way of keeping everything in his head at once.

The situation changed when Lavoisier coined neologisms that enabled him to group together whole classes of reactions so that the consequences of his proposals could be systematically explored. We are inclined to think that Lavoisier's pride in his system of chemical notation and the emphasis on nomenclature that dominates the *Traité Elementaire de Chimie* (1789) is misplaced, that a great scientific advance could not hang on anything so trivial. But Lavoisier's changes were not trivial. First for himself, and, subsequently, for his interlocutors, they fulfilled the function of making a complex body of experimental results surveyable and so enabled him to deal with classes of experiments wholesale, pinpointing the ways in which he could evade threatened inconsistencies and in which his opponents were trapped. It is not that the claims of his rivals were untranslatable into Lavoisier's language. The principal advocates of phlogiston chemistry knew how to translate Lavoisier, and he was as good as anyone at formulating his conclusions in "phlogistic" terms. But whereas they groped through a maze of separate results about "airs" and "earths," Lavoisier was able to canvas directly the possible proposals for phlogistic analyses of reduction reactions and to show systematically how they all led to trouble.

My second example is drawn from one of the rhetorically richest texts in the history of science, Darwin's *Origin*. In making his problem-solving strategies evident, Darwin uses analogies that have become famous (the analogy between natural and artificial selection, the comparison of the fossil record to a defaced book, the thousand wedges, and so forth). But rather than focusing on these devices and the cognitive roles they play, let us look at a more general feature of the *Origin*: Darwin's use of the doctrine of "special creation" as a foil.

This is a brilliant use of a common rhetorical strategy—which I employed in a humdrum fashion in the beginning of this essay—one that might justly be called "inventing the opposition."

In one obvious sense, there was no such thing as the doctrine of special creation in mid-nineteenth-century British natural history. Numerous geologists, botanists, zoologists, anatomists, and physiologists pursued a large number of questions that Darwin would draw together as the field for evolutionary biology. Virtually all of these investigators took for granted that species had been separately created, and this sometimes affected the kinds of answers that they gave. But, in exploring comparative morphology, biogeography, and embryology, Darwin systematically investigated the ways in which the phenomena could be understood given the assumption of separate creation. In doing so, he created a program that was a potential rival to his own, based on a thesis he was concerned to deny, and he showed, again and again, how that program would face inconsistency predicaments and explanatory-loss predicaments from which there was, apparently, no escape.

Was this unfair to his opponents? Hardly. Although nobody had undertaken the rival program, many Victorian scientists were deeply committed to the assumption that Darwin set at its foundation. Insofar as the questions he raised for biology could not be ignored, each of his opponents was implicitly committed to the Creationist program. But part of Darwin's own achievement was to draw these questions together as a field for scientific study, and his constitution of that field was partially advanced by the device of representing it as the battleground for two rival doctrines. The invention of "Creationism" should thus be seen as a means of rendering a complex intellectual situation intelligible, and as advancing the appreciation of Darwin's "long argument."

Two short examples must suffice; they serve as pointers to the possibility of integrating rhetorical with methodological studies. However, since I have been concerned with stressing the possibility that rhetoric may serve cognitive functions, I close by disavowing Panglossian optimism. We should not think that rhetoric in science *necessarily* seduces thought: for all that (contemporary) Creationists say, Darwin is not the original serpent. The extent to which various types of rhetorical strategy prove cognitively valuable remains an open question. Whether we focus on the institutionalized devices that govern the publication of scientific texts or on the original tropes that innovative scientists employ to obtain a hearing for their ideas, the ultimate task is to understand the ways in which certain kinds of linguistic interactions promote or detract from the community's recognition of explanatory truths. The task is not to forswear the rhetorical richness of language, but to attain at last what the serpent promised—knowledge of good and evil.

NOTES

1. Here I bypass the obvious and important concerns about how to divide processes into types. I simply suppose that a particular chain of reasoning can be seen as involving certain dispositions that could be activated on other occasions, and that our assessment proceeds by considering the general effects of these dispositions.

2. Note that this conclusion depends on the organization of the community and on the pressures for diversification within it. If, as I believe, scientists have an interest in pursuing underrepresented ideas, then rhetorical excesses may be far less harmful than is usually supposed. See Kitcher (1990a).

REFERENCES

Bloor, D. 1991. *Knowledge and Social Imagery,* 2d ed. Chicago: University of Chicago Press.

Collins, H. 1985. *Changing Order: Replication and Induction in Scientific Practice.* London: Sage Publications.

Darwin, C. [1859] 1964. *On the Origin of Species: A Facsimile of the First Edition.* Cambridge, Mass.: Harvard University Press.

Engler, R. 1987. "Misrepresentation in Biomedical Research." *The New England Journal of Medicine.*

Hardy, G. H. 1940. *A Mathematician's Apology.* Oxford: Oxford University Press.

Holmes, L. 1985. *Lavoisier and the Chemistry of Life: An Exploration of Scientific Creativity.* Madison: University of Wisconsin Press.

Kitcher, P. 1990a. "The Division of Cognitive Labor." *Journal of Philosophy.* 87: 5–22.

———. 1990b. "Persuasion." In M. Pera and W. Shea, eds., *Persuading Science.* New York: Science History Publication, pp. 3–27.

———. 1993. *The Advancement of Science: Science without Legend, Objectivity without Illusions.* Oxford: Oxford University Press.

Latour, B. 1987. *Science in Action: How to Follow Scientists and Engineers Through Society.* Cambridge, Mass.: Harvard University Press.

Laudan, L. 1984. *Science and Values: An Essay on the Aims of Science and Their Role in Scientific Debate.* Berkeley and Los Angeles: University of California Press.

Lavoisier, A. 1789. *Traité Elementaire de Chimie.* Paris: Cuchet.

Merton, R. 1973. *The Sociology of Science.* Chicago: University of Chicago Press.

Nature. 1991. Vol. 351.

Russell, B., and A. N. Whitehead. 1925. *Principia Mathematica,* 1st ed. Cambridge, England: Cambridge University Press.

Shapin, S., and S. Schaffer. 1985. *Leviathan and the Air-Pump: Hobbes, Boyle, and the Experimental Life.* Princeton: Princeton University Press.

Solomon, M. 1992. "Scientific Rationality and Human Reasoning." *Philosophy of Science:* 439–55.

———. 1994. "Social Empiricism." *Nous* 28: 325–42.

Comment

Merrilee H. Salmon

Department of History and Philosophy of Science, University of Pittsburgh

Philip Kitcher sketches a *via media* between the traditional view that science is all substance and the postmodern view that science is all surface. In the *via media*, the medium is neither the message nor the transparent carrier of the message. Rhetoric is not everything, but it matters. To become a part of science, descriptions of "scientific" activities and outcomes must be cast in a rhetorical style appropriate to scientific reasoning. No scientific language, not even a formal mathematical language, can do its work without rhetoric. Euclid may have looked on beauty bare, but in the telling of it some rhetorical drapery was added. Similarly with the proofs of formal logic. As the author of the *Port Royal Logic* put it, "If any man is unable to detect by the light of reason alone the invalidity of an argument, then he is probably incapable of understanding the rules by which we judge whether an argument is valid—and still less able to apply those rules" (Arnauld 1964, 175). Kitcher's point about the need for gas in mathematics is well taken. The gas is not just hot air. Like a whiff of oxygen at high altitudes, it enables us to see and think more clearly, to understand a proof rather than merely follow it.

In a limited, local way those of us who try to cross disciplinary boundaries know only too well that science is not a rhetoric-free zone. An argument that will persuade psychologists might not move biologists unless it is recast in their own terms, and vice versa. As archaeologist W. Adams and philosopher E. Adams acknowledge in their recent book on typology (1991), even brothers who share similar educational backgrounds and concerns find difficulty in communicating across disciplines. Temporal distances impose similar rhetorical barriers. Historians of science are frequently asked why some arguments worked and others failed when, from a contemporary point of view, both sorts of arguments seem equally good or bad.

Kitcher's concern, however, is not with the barriers to cognitive functions that rhetoric can impose, but with the cognitive functions it can serve. Moreover, he wants to study global rather than local features. He is interested in those rhetorical forms that are conducive to good science, understood in the broadest way. He believes that these forms are worthy of study because they promote epistemically valuable ends that are commonly recognized and shared. More specifically, he says that these forms promote *explanatory truth*, which he glosses as "true statements that reveal the structure of nature." One could object that this epistemic goal can be described as "commonly shared" only because it is stated vaguely enough to be interpreted differently by those on opposite sides of the instrumentalist-realist controversy. Nevertheless, agreement about a vaguely described goal may suffice to allow opposing parties to adopt the same conventions in presenting scientific results.

Which rhetorical forms lead to recognition of true statements that reveal the structure of nature? Kitcher finds them exemplified in conversational exchanges among scientists and in the public scientific texts that arise from such exchanges. He discusses rhetorical features of two sorts of scientific texts: (i) contemporary reports on new experimental findings; and (ii) articles or books that offer a novel and significant theoretical perspective. Since the feats of rhetoric employed to convince readers of the value of novel theoretic perspectives can be stunningly creative and diverse, Kitcher is to be commended for organizing these rhetorical strategies into just two functional and interrelated categories: those designed to overcome inconsistencies and those designed to avoid explanatory losses.

In this brief commentary, I focus on the everyday "invisible" dead rhetoric of research reports. Kitcher asks whether a different set of conventions would improve our chances of detecting experimental fraud. I wonder also whether other conventions would improve our recognition of incompetence.

Reports on the outcome of experiments are staple items of normal science. Each nugget is added to the golden pile, although occasionally an inferior nugget must be removed so that the new one can take its place. As Kitcher points out, besides reporting the outcome of experiments, the authors of reports try to persuade referees and editors that their experiments are competently conducted. To do this, experimenters mention the ways in which the experiment could fall short of standard practice. They then explain how pitfalls were avoided, thus anticipating and disarming criticisms. The device of weakening the force of a critic's objections by anticipating them is a standard debating technique that also figures importantly in the reasoning of ordinary conversations (Salmon and Zeitz, forthcoming). Anticipating criticisms of one's own position also helps to maintain a conversation by letting the partner know that concerns are recognized and shared.

Scientific competence is questioned when a referee is aware of some perturbing influences that the authors do not mention. Given the atmosphere of presumed trust among scientists, however, fraudulent concealment of a perturbation is rarely suspected. Indeed, a competent fraud would surely sin by commission rather than by omission, figuring that the referee would detect the omission, but that a lie would not be challenged.

The norms of scientific conduct require the research group to police itself, and to avoid or acknowledge any situations that would invalidate or weaken the experiment or throw its outcome into question. The rhetoric of the report codifies this practice, allowing the scientist or the research group to act as both defense attorney and prosecutor for the case. But the implicit analogy with a court of law is defective because in legal disputes each side deploys its skills to conceal its own flaws while trying to detect the opponent's. Many familiar rhetorical devices serve this end with the result that each side finds weaknesses revealed that it would not have anticipated prior to the adversarial treatment. Although the *via media* recognizes that science cannot be a rhetoric-free zone, it probably would balk at importing wholesale the adversarial rhetoric of the law courts. Scientists would protest the waste, the expense, and the distastefulness (but see Rescher 1977).

The image with which Kitcher's essay begins suggests a device that avoids the strong adversarial stance of a court of law, but achieves some of the same benefits: the ploy of the devil's advocate. *Paradise Lost* reminds us that if Eve and Adam had not yielded to temptation, not only would their lives have been easier, but their deaths would have assured their entry into the community of saints. Because Adam and Eve sinned, their heirs can achieve beatitude only by overcoming the effects of original sin and dying in a state of grace. Before a person can be judged a saint in the Roman Catholic Church, his or her life is subjected to close examination. Traditionally, this task was undertaken by a team of priests whose labor was divided between those who sought evidence of sanctity and the devil's advocate, whose job was to uncover any defect that would impede the candidate's entry into heaven. Thus although the devil's advocate was the adversary in the case, he acted as part of a unit whose goal was to canonize those worthy to be called saints. By borrowing the rhetorical moves of the devil's advocate, that is, by requiring a member of the research team to adopt an adversarial stance, to ask questions designed to uncover error or fraud, to use "inside information" to pursue weaknesses or objectionable features of a theory, and by incorporating some of this process into the research report, science might better avoid incompetency and fraud in its ranks.

In contrast to the research report, the presentation of a novel theory has less need for the rhetoric of the internal adversary since the defenders of the old theory will carefully scrutinize the new. Creators of new theories try to locate all troubles in the old theory, and to show that problems can be avoided by adopting the new. Since defenders of the old theory have a strong interest in maintaining it, they do not accept such claims without a struggle. Thus, proponents of new theories are relieved to some extent of the burden of finding flaws in their own theories. They have a well-defined opposition—or they create one, as Kitcher tells us that Darwin did—with more at stake than the dissatisfaction of a grumpy referee.

Kitcher's recognition and analysis of the cooperative or shared nature of scientific *reasoning* deserves special notice. While many writers comment on the shared nature of scientific activity, few recognize that reasoning itself is a shared activity. Questions and challenges within the group can redirect the flow of arguments; different interpretations of analogies and examples, or outright disagreements, can lead members down different paths of thought. The overall result of the group activity is not so much an assembly of premises that support some conclusion, but rather a compromise position forged from a guiding idea, challenges, attacks, false starts, insights that need to be supported, and rejected alternatives. Research reports need to represent this often inefficient process efficiently, but they should incorporate enough of the process to foreclose concealment of incompetence or fraud.

REFERENCES

Adams, W., and E. Adams. 1991. *Archaeological Typology and Practical Reality: A Dialectical Approach to Artifact Classification and Sorting*. Cambridge, England: Cambridge University Press.

Arnauld, A. 1964. *The Art of Thinking: The Port Royal Logic*. Translated by J. Dickoff and P. James. Indianapolis: Bobbs-Merrill Company.

Rescher, N. 1977. *Dialectics: A Controversy-Oriented Approach to the Theory of Knowledge*. Albany: SUNY Press.

Salmon, M., and C. Zeitz. Forthcoming. "A Method for Analyzing Conversation Reasoning." *Informal Logic*.

4

How to Tell the Dancer from the Dance

Limits and Proportions in Argument About the Nature of Science

J. E. McGuire
Department of History and Philosophy of Science, University of Pittsburgh

Trevor Melia
Department of Communication, University of Pittsburgh

In one of those apercus for which he is justly famous, K. Burke insists that we are, all of us, "rotten with perfection." The seeming incongruity of the aphorism is somewhat relieved when we learn that the "perfection" alluded to is not that of a Hollywood beauty gyrating to the rhythm of Ravel's *Bolero*. What Burke has in mind is the Greek notion of perfection as completeness, particularly symbolic completeness. That is the sort of perfection that Russell and Whitehead aspired to when, after 362 pages of manipulation of the arcane notations of symbolic logic, they blandly announce in *Principia Mathematica* (1925) that they have shown that one plus one really does equal two. We do, in fact, get carried away by the internal logic of our symbol systems. Even the dedicated empiricist who starts determined to let the "facts speak for themselves" finishes usually by exemplifying Coleridge's dictum that "language itself, does as it were *think* for us" (1907, l.63). Little wonder that postmodernists are driven to conclude that we are, to use F. Jameson's resonant phrase, incarcerated in the "prison house of language."

It is a not-so-fine point in Burke scholarship whether the sage of Andover is, or is not, a proto postmodernist. Burke, after all, insists that we are not only the symbol-*using* animal but also the symbol-*used* animal. Before Derrida announced "il n' y a pas dehors text," Burke averred that "things are the signs of words" (1961, 361). And, of course, there is the famous doctrine of "terministic screens," "Men seek for vocabularies that will be faithful *reflections* of reality. To this end they must develop vocabularies that are *selections* of reality. And any selection of reality must, in certain circumstances, function as a *deflection* of reality" (1945, 59).

Still, if Burke is to be regarded as a Jamesonian jailor it must be allowed that for him the dimensions of the "prison house of language" are Protean. Meanwhile, there is in that troublesome and much overlooked caveat "in certain circumstances" which implies that in some other circumstances language does not "function as a deflection of reality" (ibid.). That is, Burke allows for "minimal realism" (McGuire and Melia 1989). At first blush, it is surprising to find Burke, so much an architect of the "linguistic turn," seeming to resist the currently fashionable tendency to reduce everything to language and text. Nevertheless, if anything is more basic for Burke than his insistence on the primacy of human symbolicity, it is his resistance to hegemonic impulses of any kind. Burke recognizes that human "rottenness with perfection" is most ubiquitously manifested in attempts to establish the hegemony of a single vocabulary or "voice." These "essentializing" thrusts currently most likely to be directed against science are also deeply antithetical to the very idea of argument.

The attempt to show that the bewildering multiplicity of things is but the manifestation of a single overarching entity to which, properly construed, all things can be reduced, has been continuous and continuously productive through the ages. Monotheism, which finds its most celebrated secular variant in Platonism, has been in combat with polytheism from its beginning. On the sacred front, our times are more likely to be characterized by clashes between monotheisms. "My God is the only God" or somewhat more modestly "My God is more powerful than thy God." The most powerful secular variant for modernity has been, and in many ways still is, *science*. In that, *almost* perfect, scheme, as the Vienna school positivists reminded us, the chaotic complexities of social life are explainable by reference to the statistical regularities of aggregated individual psychologies. Individual psychology, in its turn, can be made transparent by a biology which is reducible to an even more disciplined organic chemistry. The complex molecules of chemistry can be entirely accounted for by physics. Meanwhile, the multiplicity of phenomena known to physics are really manifestations of the operations of a few fundamental forces, according to the grand unification theorists, coalesced out of a single force that prevailed for a singular instant at the beginning of time, "In the beginning was the" Force, and to complete the *bouleversement,* "may the Force be with you." But almost all well-bred humanists, and all poets, know that "in the beginning was the Logos," that is, the *word.* So Burke, who, as Marianne Moore reminds us, is "first of all a poet," leaning against the prevailing perfection of his times offers as a corrective to materialistic scientism in his own linguistic dramatism.

It would be a mistake, however, to read Burke's dramatistic critique of scientism as rendering all science *"sub specie rhetoricae"* (Melia 1989, 56). That perfecting posture, most recently adopted by A. Gross, flies in the face of Burke's steady insistence on the symbol-using creatures' *animality,* subject to ecological rules and facing various orders of *recalcitrance* as we attempt to operationalize our symbolic representations of reality. If Burke can be reduced to a few central injunctions, they are: (1) *resist the intrinsic lure of language, and therefore the natural tendency of human psychology, to essentialize;* (2) *seek always to acknowledge and proportionalize primary constituents,* (3) *look for signs of recalcitrance when these injunctions are ignored.* Burke continually warns against the tendency of Occam's Razor to oversimplify; for as he observes several times in his Dialogue in Heaven, "it is more complicated than that" (1961, 274). His *Grammar of Motives* (1945) is a systematic exposition of the recalcitrance faced by various philosophies as they seek to essentialize one or another of the terms of the ubiquitous "dramatistic" pentad. Burke's dramatism, unlike that of I. Goffman's, is not simply a mobilization of the theater metaphor, but is rather a disciplined attempt to deploy an adequate, nonreductive, critical vocabulary. Insofar as humans ACT whereas things merely MOVE, Burke insists that the adequate discussion of any human undertaking will require attention to all aspects of ACT, hence, the pentad: Act, Scene, Agent, Agency, and Purpose (later expanded to a hexad by the addition of Attitude). Burke recognizes that all action presupposes motion and he is respectful of scientific analyses of Scene (Nature). But while the scene is in one sense the given, it must also be taken. The Scene must be Seen. And the notion of terministic screens implies that seeing is ACTIVELY appropriating our surroundings. Burke insists also that the relations between the key terms in his critical vocabulary (the ratios) be taken into account. Most particularly he emphasizes the importance of the Scene-Act ratio and insofar as the flexibilities of Scene allow, or rather insist, that the human context be treated, science must be seen as a socially embedded Act. The fact that words are not things leads to all sorts of embarrassments in our symbolic attempts to appropriate the world. But, we argue, those difficulties hardly warrant the flight from reality implicit in the attempt to see everything *sub specie rhetoricae*. Rather we should expect, and even welcome, the ambiguities consequent on our symbolic situation. Like Burke, we hope to find in linguistic incongruities, not despair, but perspective—all the while favoring the proportionalities of irony over the crude reductions of metonymy.

In this essay we illustrate the Burkean injunctions as they apply to contemporary argument about the nature of science. More specifically we reassert our resistance to three powerful hegemonic thrusts. The first, what may be called "naive realism," is perhaps best exemplified by S. Johnson's celebrated proposal to refute Berkelian idealism by kicking a stone. Because Gross has mistakenly assimilated us to the position, we are primarily concerned with clarifying our posture vis-à-vis naive realism. We also limn out our position on two other hegemonic thrusts which will be the subject of detailed analysis in forthcoming papers. The second hegemonic thrust carried out by all the variants on Bloor's Strong Programme seeks perfection by reducing science to sociology. The third, most manifest in Latour and Woolgar (1986), by its emphasis on inscriptional devices, reduces science to text. We precede these discussions, however, by a description of our position which attempts to resist the tendency, exemplified in varying degrees by each of the mentioned caricatures, to treat science as a system of *representation* rather than a set of situated and constrained *practices*.

I

The principal norms of scientific practice are often claimed to be autonomous and invariant over time. Sometimes this is modified modestly to the view that this is so at least since the seventeenth century (Stump 1991; Doppelt 1983, 123–24). Those who argue for the fixity of scientific methodology fear the threat of nonfoundationalism and hence the bogie of relativism (Worrall 1988). Specifically, they fear that, if scientific rationality changes and if principles of evaluation are not fixed, we have no Archimedian point from which to judge scientific progress. But if this is so, they argue, relativism ensues. Moreover, they maintain that relativism can be countered only by securing fixed standards for scientific practice which rest on secure epistemic foundations. We consider the recalcitrances involved in scientific practice by taking a stand in the ongoing debate between the foundationalists and the nonfoundationalists concerning the nature of scientific knowledge. In this connection we once again consider Gross's (1990) view that scientific theorizing is (1) a species of rhetorical invention; (2) is to be understood without remainder as a system of linguistic representation; and (3) is completely relativized to time, place, and context. Gross is, of course, an antifoundationalist who wishes to clear the field for rhetorical practice. We deal with this aspect of his position in McGuire and Melia (1991). In the present essay, we use his views as a springboard for a consideration of the implications of fallibilism (or epistemic nonfoundationalism) for the rhetoric of science.

If science is seen as a socially embedded act, the human context of scientific inquiry becomes central. According to the perspective we advocate, science is a form of praxis, that is, a particular form of human activity with identifiable characteristics and embodying definite methodological aims and goals. Since, however, all forms of human activity are situated in symbolic contexts, scientific praxis is always and already a form of symbolic action involving both techniques of intervention and representation. This is important; science is at one and the same time an enterprise that intervenes designfully in the world and represents the world from various perspectives. These forms of activity are both constitutively symbolic in as much as intervention and representation are acts embedded in meaning, intention, purpose, and design. Thus, from our perspective, science is symbolic praxis expressed in a set of material practices governed by the methodological norms of those practices and embodied in the institutional framework of its self-generated contexts.

A key term here is practice. If, as we suggest, science is seen as a particular form of human activity or collective praxis, it can fruitfully be understood through the practices that embody this activity. By a scientific practice we mean a determinate form of human activity that at once expresses a style of reasoning and discloses an identifiable pattern of action. Scientific practices, therefore, constitute many and various forms of human activity ranging across a spectrum from the purely symbolic to the discursive, the experimental, and the technical. They include the local and micropractices of laboratories as well as the practices embedded in the global networks of scientific communication. Moreover, the practices of science are interrelated and underwrite in their totality the possibilities open to scientific inquiry at any particular point in time. Nothing is inviolate about scientific practices, or about the norms and aims which drive their exercise. Over time any or all of them can change and modify within the ongoing adaptive dynamic of the scientific enterprise.

If scientific praxis possesses a recognizable symbolic structure directing a self-correcting activity which carries with it its own ever-changing norms, aims, and goals, it already possesses rhetorical dimensions and invites rhetorical analysis. More particularly, it invites modes of analysis that are sensitive not only to the ways in which rhetorical strategies and devices function in the discourse of science, but also to the ways in which larger cultural formations may or may not impinge on the practices of science. Furthermore, a rhetorical orientation calls attention to those sites of contestation which are frequently the midwife of scientific innovation and productive of new patterns of scientific knowledge. It is here, moreover, that rhetorical analysis can help to illuminate how key discursive components (such as those of the Act-Scene ratio) generate various forms of recalcitrance and foster ambiguity consequent upon their symbolic situation. The symbolicity of scientific praxis generates in itself forms of rhetorical presentation.

According to the "interactive" model of scientific practice, choice of aims, standards, and methods involves reference to the actualities of scientific practice. This means that the methods and aims which are entrenched in the activity of making science emerge from the realizable results achieved by the practices of science itself. In our view, norms, aims, and standards cannot be set apart from, and appraised independently of, the practices of science. To think otherwise is to essentialize scientific methodology as an end in itself, and to detach the norms and aims which guide science from the empirical results established and realized through its methods of inquiry. This does not mean, however, that the scientific enterprise is condemned to move in a holistic circle. It means, rather, that the actualities of scientific practice over time set *internal* constraints on the sorts of criteria that are applicable and realizable in the making of science. To realize that the standards and norms which connect the means and aims of scientific practice change, and have changed, confirms two conclusions. First, the hegemonic thrust which seeks to establish a fixed and invariant methodology for science is on shaky ground. Second, we cannot overlook the fact that burgeoning information regarding the constitution of the physical world which science continually amasses provides a significant constraint on the sorts of aims, norms, and methods which science continues to project. These conclusions, in our view, provide no grounds for the hegemonic thrust of unbridled realism anymore than they underwrite the view that science is constitutive of its modes of inquiry independently of its objects of inquiry. We have an important moral here for our understanding of the rhetorical dimensions of scientific practice.

II

We turn to a general discussion of these claims. Specifically, we show that a rejection of foundationalisms, such as metaphysical realism, coherentism, and transcendentalism, neither denies the possibility of objective knowledge nor leads to relativism. Since the achievements of the scientific revolution of the seventeenth century, we have increasingly justified the idea that we have learned progressively more about the physical world, generation after generation. This fact is clearly evident in the ever-increasing control we realize over our physical environment and in the astounding technologies we continue to produce in the wake of our deepening scientific knowledge. The scientific enterprise is unique. In its modern form it has come into being only over the last three centuries. To anyone familiar with the history of the sciences in the modern period, it is evident that there have been both failures and successes. Evident also is the fact that methods and aims of various sciences differ among themselves at a given time as well as change over time within their separate domains. Clearly no hegemonic method fits all the sciences at any time any more than an invariant set of aims guide them through time. It is folly, then, to suppose that the dictum "whenever possible subject your theory to severe tests" is a procedural rule exemplified in the actual practice of all good science. If it were true, it would be vacuous. But it is not true since it is not evident in the methods of many of the sciences past and present.

If the sciences are not embarked on the quest of essentializing a hegemonic method, what features serve to unify the sciences as an identifiable enterprise? If the sciences are as heterogeneous as we claim, how can we view scientific inquiry at different periods of time as involved in a common enterprise? First, science creates an ideology of the great exemplars of its past achievements such as Newtonian theory, Darwinian evolution, Einstein's theories of relativity, and the esoterica of the new quantum theory. This ideology of progress is intentionally legitimating and celebrates the great innovations, the classical experiments, and the realization of powerful new techniques. Second, these great moments generate through the ideology a canonical understanding of its triumphal achievements which the scientific community can identify as belonging to its common enterprise. Last, these canons of achievement, though part of the ideology, are not merely fanciful; they play a legitimating role in the ongoing activity of the scientific enterprise. So although the aims and methods of the sciences are changing, a clear sense of these canonical achievements is preserved in the articulation of the methods, norms, and reasons for theory choice in the various scientific disciplines.

We turn now to the foundationalist and nonfoundationalist debates. Consider first the clear and uncompromising position put forth by Worrall. Worrall defends fixed and invariant norms of theory appraisal, "If no principles of evaluation stay fixed, then there is no 'objective viewpoint' from which we can show that progress occurred—However this is dressed up, it is relativism" (1988, 274). Note that Worrall cobbles together the notion of "objective knowledge" with a god's-eye view of the aims and norms of science: You cannot have one *without* the other, and only if you have one *with* the other do you have a clear alternative to relativism. In a direct response to Worrall, L. Laudan has correctly insisted that Worrall misunderstands the threat from relativism. In Laudan's view the relativist's charge is serious whether the methods of science are one or many, constant or changing (Worrall 1989). What the relativist demands is a nonquestion-begging rationale for those norms of theory appraisal which we employ at *any* given time in the scientific enterprise.

Laudan's response points in the right direction. Certainly relativism makes justificatory demands on any methodology, and it certainly claims that criteria of justification are tied to time, place, and context. It does not follow, however, that objective knowledge is impossible.

In part, this is a matter of what we take objective scientific knowledge to be. If it is construed in Platonic terms as a body of invariant propositions, then the relativist is correct to demand a justification of each standard against the vicissitudes of scientific change. Alternatively, if scientific knowledge is construed as that set of theories and techniques accepted by practitioners in certain domains of scientific inquiry, the question of justification differs. On this understanding of scientific knowledge, we argue, the question of justification becomes a matter of being able to show that certain standards and methods are better than alternatives as guides to the production of scientific knowledge.

Now it seems ironic to insist that a plausible philosophy of science, and indeed a plausible rhetoric of science, base themselves on the actualities of scientific practice. Unfortunately, that insistence too often goes unheeded. Contrary to what Worrall supposes, the desideratum of "objective knowledge" is not tied intrinsically to a god's-eye perspective on scientific methodology. Nor is science constructed according to the techniques of rhetorical invention, a set of representations divorced from the "facts" of science as Gross (1990) supposes. Ironically, both Worrall and Gross misrepresent the practice of science: Worrall because he fears relativism, Gross because he embraces relativism.

Let us begin with the relationship between the practice of science and the philosophical justification of science. It is foolhardy to press upon science an epistemological perspective that is alien to its practices. A perspective is needed that takes the practice of science as primary, and which finds the justificatory reasons for the norms and aims of that practice within the scientific enterprise. That is, we support an epistemology which rests squarely on the explanatory strategies involved in the actual practices of science. This does not mean, however, that we are merely grounding ourselves in the actualities of science and fitting the tools of our analysis to the protocols of scientific practice itself.

In McGuire and Melia (1991) we argue that existence claims, regarding the constitution of the physical world, are not to be confused with accounts of truth. Specifically, we argue that questions of the justification of the sort of minimal realism we espouse should not be confounded with issues of truth. To extend that perspective, the sort of justification that the nonfoundationalist proposes is not the justification of a claim to truth, but the claim that, within science itself, justificatory parameters can be established which promote the acceptance of a theory, given certain aims, norms, and methods.

Notice the constraints involved. First, metalevel justifications of science are constrained by the practices of science. Moreover, the practices of science, at a given time, limit the range of justificatory reasons that can be advanced from within science. This means that, since the aims and norms of science change through time, the acceptable reasons available for justification are constrained both at a time and by change over time. For example, the justificatory resources of the Newtonian research tradition differ from those available within the Einsteinian, and both traditions have evolved different aims and methods over time. The Newtonian tradition aims to establish ontologies that invoke centrally directed forces operating at a distance, and to force laws expressed mathematically in terms of inverse-square relationships. Nevertheless, within that tradition is a move from an inductive to a hypothetico-deductive methodology of justification. Furthermore, in relativity we are constrained to treat mass and energy as equivalent and interchangeable, and in the new quantum physics particles are convertible into waves, and vice versa.

Against this background consider the question of the relationship between "objective knowledge" and fallibilism. If scientific practice is the ultimate arbiter of what counts as success in science, it sets constraints on the available standards by which a piece of science is to be judged. Moreover, since scientific practice does not generate certain knowledge, the claim that science must rest ultimately on self-justifying foundations is an illusion. These two observations support the claim that if objective scientific knowledge is possible, it must be settled within the fallibilist structure of the scientific enterprise.

Now this perspective meets a timeworn difficulty. Put simply, how can scientific success in establishing cognitive control over our physical environment be justified in terms of itself? Not only must we justify scientific success within the framework of scientific practice, but we must do so in a way that is patently circular. Our two main answers to this difficulty emphasize the importance of understanding the sorts of constraints inherent in science.

The first answer says, in effect, what is the problem? If epistemic foundationalism is rejected, so too is any recourse to a priori metastandards or stipulated conventions. This means that either we (1) accept circularity; or (2) show that it does not obtain in an objectionable form. On either view, one accepts that the canons of justification can only be established from *within* science. But on the second response the sting of circularity can be mitigated. Although norms, aims and methods interconnect, they are not interdependent. Thus, not only do we have independent criteria for implementing rules and norms in a particular scientific domain, but often we have different sets of criteria to which appeal can be made. Thus, instead of an invariant set of holistic standards to which appeal is made, in fact, a multileveled, multifaceted range of justificatory criteria is available for assessments from within a scientific domain. These criteria cannot be called into question *all at once* (as those who accept fixed methodology suppose), but any one of them can be called into question sooner or later. Thus, criteria of theory choice and of theory acceptance can be different at each stage in scientific inquiry, and each is revisable as inquiry moves diachronically through successive contexts.

The second main answer is found in the work of Dewey (1938), Rescher (1977), and Laudan (1990). This answer argues that the issue is not one of vitiating circularity, but of recognizing the symbiotic nature of scientific practice. That practice, as we are characterizing it, is not cumulative, linear, and sequential. If it were, the circularity would be evident and vicious. Moreover, the changing norms and aims of science function within the process of inquiry as fallible posits or conjectures. In Laudan's view, they are best construed as hypothetical imperatives, linking means to ends—imperatives which make factual reference to the sorts of connections which knit together the means-ends continuum. He also insists that strong constraints on the aims of science are permitted in scientific practice. They must be realizable; otherwise, there is no means to promote their realizability. This is true not only for the internal norms and rules of science, but also for procedural rules of methodology. The latter are "abstracted" from the actualities of practice and are thus an expression of the canonical ideology of science itself. A case in point is P. W. Bridgman's operationalism. Noting that special relativity is a consequence of asking "how can we determine simultaneity" ([1927] 1961, 8), he raises Einstein's operational procedure to a canonical method. Typically, however, he hopes for too much when he avers that operationalism will forestall the "danger of having to revise our attitude toward nature" (ibid., 4).

III

We can now consider "objective knowledge," the charge of relativism, and the claim that the physical world constitutes an external constraint on the changing practices of science. What conception of knowledge does the scientific enterprise support? Science, as indeed any form of human activity, dwells in its practices. The practices which make science what it is are constitutive of the sorts of directed actions characteristic of that specialized form of human activity. Here the term "contitutive" refers to the fact that the sorts of practices—such as theoretical, experimental, and technical—which get exercised in a line of inquiry are formative of the knowledge and understanding that is realized. In key part, scientific knowledge "exhibits" itself in various forms of quantitative and technical control over nature. Accordingly, an unquestionable relationship exists between knowledge and the exercise of the practices which it underwrites.

Indeed, the relationship is symbiotic and interactive. Although scientific knowledge expresses itself through the exercise of its practices, at the same time these practices are driven by that knowledge. Science cannot, of course, transcend the networks of its practices. What science understands, then, must arise from those practices and be made intelligible and useful through the aims and norms which guide them. This is not to say that scientific practice avoids appeal to "superempirical" norms and aims, such as explanatory fertility, coherence, or the unifying power of theory. However, these second-order cognitive virtues are only manifested over time through the career of theories and their associated practices. However, note that these virtues ride on the back of those successful practices which produce dividends and possess a realized track record. And there is no question here of essentializing a procedural norm ex post facto, for example, "Always seek theories of maximal fertility," for there is no way of establishing at a given time whether a procedural norm will continue to have a basis in future scientific practice. Clearly, objectivity in any area of practice requires methods for accepting or not accepting the observations and judgements of qualified practitioners. Such methods, however, are set by the discipline in question: There is no such thing as objectivity in general.

Relativism is a notoriously ambiguous term. Gross equivocates on its descriptive and normative uses. He moves unwittingly from the benign observation that there is at different times diversity of opinion in contingently different scientific contexts, to the conclusion that these differing contexts warrant prescriptively different norms and aims of scientific practice. This will not do; descriptive relativism is logically independent of normative relativism. From contingent facts nothing normative follows. But equally, Worrall's position is untenable. We do not need fixed norms to justify continuing practice and to defeat the relativist. Not only do Gross and Worrall misrepresent the nature of scientific practice, but they obscure the fact that that practice gives little comfort to the relativist. Undoubtedly, beliefs and practices generated within one perspective often migrate beyond it and become adapted by another perspective. This is especially so in the sciences. Think of the creative symbiosis of physics and biology that unraveled the DNA molecule, or the fertile marriage between cosmology and particle physics. Or think of the key commitments of the Newtonian research program: inverse-squared distance laws and central forces acting at a distance. This program survived largely intact into late nineteenth-century physics and is still adapted into nonquantized and nonrelativized domains. Accordingly, norms and practices adapt to different problems in different contexts at different times. There is a continual interplay between aims, norms, and methods as inquiry proceeds. Nothing stays completely fixed; nevertheless, notable continuity exists in the practices of science over time. To defeat the relativist, we need only point out that, within the ongoing continuum of scientific inquiry, exist rules and norms capable of justifying the acceptance of a given theory over its rivals, and this means having reasons to hold that a given theory is the best now available for pursuing the business of controlling our physical environment.

Clearly, then, essentializing methodologies falsify the nature of the scientific enterprise; but they also saddle the procedures of science, through their perfecting tendencies, with artificial constraints. If an inadequate philosophy of science fails to recognize this, a rhetoric of science which does likewise makes itself irrelevant to its task. Gross's disposition to portray science as a system of representation, as a text to be analyzed *sub specia rhetoricae,* is a case in point. Science is, of course, in virtue of its symbolic resources a representational enterprise. But it is also in virtue of these same resources an instrument of intervention in virtue of its situated practices. Neither its representational nor its interventionist resources should be essentialized; both should be proportionalized. Thus, far from advocating an essentializing rhetoric of representation in the manner of Gross, we propose a proportionalizing rhetoric which reflects the proportionalizing strategies of scientific fallibilism.

But what is it to say that science intervenes in the physical world in virtue of its situated practices? Does this not invoke "the real" and reduce science to the hegemonic thrust of realism? And does it not invite the specter of foundationalism? Furthermore, does it not qualify our view that science operates with internal constraints of its own making? If we advocate a proportionalizing balance between representation and intervention, equally we refrain from essentializing either the methods of science or their referent, the physical world. Nevertheless, we defend a version of minimal realism. As we have argued elsewhere (McGuire and Melia 1991), science can and does fix nonlinguistic referents to the physical world, a task that can be performed only through the mediation of its situated practices.

We now turn to the notion that the actual presence of the physical world constitutes in itself a form of external constraint on the practices of science. Natural things not only manifest striking proportions, they also bear indelible limits. The inherent strength of materials disallows the construction of objects beyond a definite size; the Carnot cycle forbids perpetual motion; and there are limits to the depth of descent into the oceans without compensation for those limits. There are also, although few in number, the universal constants, such as G, the constant of universal gravitation; C, the speed of light inherent in relativity; and h, Planck's constant, basic to quantum theory. In Newtonian theory, for example, G functions as a constant of proportionality, having the same value in all circumstances, constant in the mutual attraction of a stone and the earth, of the earth and its moon, of the sun and a planet, of one star and another, or of two grains of sand on a beach. Each of these circumstances, in its own characteristic domain, presents limits which the practices of science cannot transcend. Always bodies which mutually attract each other, according to speeds less than C, do so in direct proportion to their masses and inversely as their distances squared. We cannot know a priori whether these are necessary limitations on practice and theory, nor can we hold that they are invariant and fixed. Einstein's cosmological speculations help to illustrate this point. In his 1917 cosmological paper he invoked the cosmic constant in order to make his model of the universe stand still. The equations of general relativity allow for either an expanding or a contracting universe, but not for one that is standing still. Only by adding the cosmic constant could Einstein provide an interpretation of the equations that allowed for neither an expanding nor a contracting universe. However, in a joint paper with DeSitter, Einstein (1932) dispensed with the constant, realizing that his equations told him that the universe is dynamic and expanding. Nevertheless, the constants obviously set limits to practice and to experimental interventions directed by scientific inquiry.

Now it is reasonable to assume that the recalcitrances confronting practice have a factual basis in the physical world. Although we abhor such vague locutions as "the world makes true" or "the world shapes causally the beliefs of the scientific community," we certainly maintain that scientific practice is "guided" by nature. Accordingly, to claim that there are nonintentional and nonrepresentational yet causal linkages with the physical world comes to this: that we are the product of, and are situated in, environments which we shape, and by which we are shaped, through the interactive capacities of our senses. Moreover, we can interact only according to abilities commensurate with the sort of scientific praxis we possess and its augmentation through instrumentation. The details of the various causal stories involved are filled in again and again by the practiced resources of the relevant investigative sciences. Indeed, the discursive practices of science have generated a tradition of establishing better and better acts of control over our natural environments, exercises which often do not involve better and better acts of representing. This alone should lead us to suspect that the epistemic arrow is pointing in the wrong direction. Criteria of scientific acceptance and justification are in need neither of transcendental epistemology nor of foundationalist realism. Scientific understanding is, to a large extent, based on its practices and on the technological possibilities which they produce. There is a *verstehen* involved here— a knowing-how—which is not captured by the notion of knowing-that. The latter is expressed linguistically and thus representationally and is essentialized by Gross in his portrayal of science and its rhetorics. In our view, the criteria mustered in the justification of a piece of science should rest on a track record of success over time, the extent that it harmonizes with other entrenched scientific beliefs, and its practical potential for further creative and resourceful inquiry. Thus, neither representation nor intervention should be privileged: An understanding of how the proportionalized balance between representation and investigative practice works in the context of scientific inquiry is needed.

The hegemonic thrust of (absolute) relativism has garnered plausibility from R. Rorty's (1979) critique of the ocular metaphor that underlies much foundationalist arguments. Rorty has little difficulty showing that our linguistic representations cannot be said to "mirror" reality. In a characteristic polar bifurcation C. Geertz draws the conclusion that "The road to discovery (lies) not in postulating forces and measuring them but in noting expressions and inspecting them" (quoted in Gross 1990, 4.) A shift to an auditory metaphor will illustrate our proportionalizing response.

In pre-industrial times the old sailing ships approaching New England's frequently fogbound shore, lacking any precise means of establishing longitude, were in constant danger of running on the rocks. Deprived of visual clues by the fog, sailors usually resorted to using sound to establish their proximity to the shore. Thus, a member of the crew would be stationed on the bow to shout from time to time into the enveloping fog. An echoed response indicated the presence of an unseen landfall. If the call were greeted by silence it was assumed that the vessel was still "at sea."

Now no one would argue that the returning echo is literally a picture of the reality from which it is reflected. Indeed if we treat the initial cry as an interrogation then the response bears a remarkable resemblance to the query. (This is the sort of embarrassment encountered in quantum theory and practice.) We can even imagine an assertive sailor crying out, "There are no rocks here!" And, if he reacted after the fashion of a naive realist upon hearing the response, "No rocks here," he might forge ahead to encounter an almost literal recalcitrance on the rocks, whereat a timid, and chastened, Jamesonian sailor might conclude that the mocking echo is testimony to our incarceration in the "prison house of language." On our view, neither response is justified. Science does not give us a god's-eye view, or even a perspectival *picture* of reality. *The answers we get from nature contain uneliminable elements of the discursive practices (symbol systems) by which we interrogate it.* That fact, among others, warrants rhetorical interest in the findings of science. Nevertheless, the returning echo is not due to the symbolic element alone as the silence that greets us when we are still "at sea" testifies. Thus, provided we do not, like the realist sailor, take the echoes literally, we can and do use them to guide our practices. And, if our practices, discursive and nondiscursive alike, are successful, it is reasonable to assume that we have made a "minimal" contact with reality.

IV

Let us now consider which rhetoric best serves the practices of science as we describe them. Obviously, one of the core beliefs that drives the practices of science is that changing norms, aims, and methods justify techniques which interact with the physical world. This in no way demands that the successes of science be justified in terms of a naive realist construal of the scientific enterprise. After all, if our portrayal of science is correct, that sort of foundationalist epistemology is wide of the mark. But so too is Gross's essentializing rhetoric which fails to accommodate the complexities of scientific inquiry.

The alternative we suggest seriously considers the proportionalizing strategies inherent in the making of science. Rhetoric must address the fact that scientific discourse is a multileveled phenomenon. It is designed for different purposes and performs various functions in the changing contexts of inquiry. For example, the vocabularies of biology have no place in the domain of particle physics anymore than the explanatory scope of the latter is appropriate in geophysics. Furthermore, a proportionalizing rhetoric is needed which reflects the proportionalizing strategies of scientific fallibilism. Often a scientific inquirer reports on an experiment to relate narratively what has happened on that particular occasion. Now for that particular experiment to make a contribution to the general beliefs held in a scientific domain, an appropriate argument framework is needed which allows the experiment to function for that epistemic purpose.

Let us look more closely at this issue since rhetoric plays a role. Whenever theories are exercised in an experimental setting they are expressed through the practices and the technologies which serve them. Just as the norms which guide practical deliberations are modified through application to changing situations, so theories, as expressed through experimental practices, are modified through adaptation to events produced and localized in the laboratory. Scientific inquiry in its laboratory setting is a constant interplay between the place where experimental events are produced, the means by which they are produced, and the significance of what is produced. While the significance of experimental practices is partly underwritten by the theory that helps to guide those actions, it is also imparted through the narrative account of what was involved in the performing of an experiment, for example, the place of the experiment, the design and reliability of the instrumentation, and the manner in which the experiment was performed.

The narrative, however, should not be mistaken for the experiment itself. Clearly, various sorts of practices are involved; that is, both linguistic representation and experimental intervention are present. The experimenter's aim is certainly to convince the relevant scientific community that what happened in the experimental setting did *in fact* occur. This involves employing linguistic techniques and strategies designed to ensure that outcome. Thus rhetorical components are indigenous to the act of experimental reporting. We would be wrong, however, to suppose that the experimental "text" alone constitutes the knowledge made. This essentializes the representative component of the experimental process, a maneuver inherent in Gross's position. It masks the fact that experimenting is *itself* a cognitive act which brings about a preconceived end. While the significance of the experimental result is conveyed through its linguistic representation, scientific understanding of the physical world is not thereby reducible to textual or verbal representation alone.

Latour and Woolgar, by emphasizing the social and textual nature of *Laboratory Life* (1986), are widely regarded as having accomplished such a reduction, but note that in his introduction to that text J. Salk, a practitioner of science, indicates some misgivings about its conclusions. Salk, whose Institute for Biological Studies was the site of the investigation, can hardly deny that scientific research is a social activity and as such can be depicted in sociological terms. No doubt the circumstances of the earlier discovery of the Salk polio vaccine would also surrender to a hegemonic description as a social process. And there were many such socially mediated attempts to discover a vaccine—most of them failures. Though he does not mention it, the fact that the recipients of the Salk vaccine did not succumb to paralyzing polio must loom large in Salk's mind. All processes may be in some sense sociological. Nevertheless, all are not equal. As we have said in McGuire and Melia (1989), "From methodological premises about how people behave we can scarcely derive ontological conclusions about what is the case in nature. This is to confound *how* knowledge is produced with what the knowledge is *about*" (p. 97).

It is true that the protocols of science have the effect of concealing the social component of the enterprise, and, where such components are admitted, they are treated as "external" and secondary, or as eventuating in mistakes. Thus Forman's (1971) study of the social conditions attending the development of quantum physics is treated as interesting but epistemically irrelevant. It is allowed that social factors in the Soviet Union led to the adoption of Lysenko's thesis of the heritability of acquired characteristics, but, of course, the thesis is also taken to be a mistake. Scientists anxious to portray themselves as "realists" inevitably invite the sociological riposte. Such is the nature of Barnes and Bloor's (1982) Strong Programme in the sociology of science. Though not themselves antirealists, by opposing one hegemonic thrust (realism) with another (sociologism), they illustrate a second type of recalcitrance. Any such essentialized depiction, no matter how plausible as a *representation*, will run into anomalies in *practice*. And those anomalies will, all too frequently, serve as the grounds for a counterthrust, which, because of the zeal of its proponents, will also be totalizing. From the Burkean point of view dramatism can serve as a therapeutic antidote to the excesses of essentialized representation. Application of the pentadic (hexadic) critique, conscientiously undertaken, clearly reveals the nonproportionality of totalizing discourse.

Essentialized accounts of science are likely, at least initially, to be received enthusiastically by practitioners of the discipline whose pretensions are thereby aggrandized. Rhetoricians may have difficulty seeing *sub specie aeternitatis,* but they are trained to view things *sub specie rhetoricae.* Bloor can count on the plausibility of the sociological view for sociologists, and we have Einstein's word that the difficulties of science are such that in order to cope scientists must believe that they are dealing with "reality." None of these essentialized accounts is wrong; all are in varying degrees inadequate. It is in the recalcitrance they meet as they encounter each other in the broader context of interdisciplinary practice that their inadequacy, at least as representation, is typically discovered.

The encounter with this type of recalcitrance can be delayed, and profitably so, by maintaining the insularity of disciplines. Thus even as philosophical realists preserve their "perfection" by ruling out social components as "external," so Gross's view *sub specie rhetoricae* would, implausibly, exclude the discovery of facts as part of the scientific enterprise. The "discovery of facts" is, of course, the part of science for which the rhetoricized view cannot account.

Even in the absence of "external" recalcitrance of the foregoing sort, the perfecting tendencies of terminologies may be expected to encounter the recalcitrance intrinsic to symbol systems generally. Ordinarily the "perfection" of a symbol system is achieved at some cost to its relevance in reference to nonlinguistic practices. Indeed, the only practices acknowledged by formal systems are those intrinsic to the system. Here, one might expect the restraints of recalcitrance to be avoidable. But as G. Frege discovered at the hands of Russell, and Russell in turn learned from K. Gödel, the attempts to articulate a symbol system that is at once complete and at the same time coherent is fraught with difficulties. Ordinary language systems normally avoid such embarrassments by the use of equivocating terms, and strategic, scarcely detectable shifts in metaphor. Nevertheless, the more a vocabulary evades the restraints of practice, and the more it achieves terminological perfection, the greater the likelihood that it will encounter an intrinsic recalcitrance.

Thus, Burke (1945) encounters the "antinomies of definition." He points out that prescientific philosophy equivocated on its pivotal term "substance," meaning by it both "that which a thing essentially is" and (etymologically) "that upon which a thing stands." It seeks thus to provide the context for its own text. This ubiquitous problem, common to all theories that reduce everything to text, is perhaps best captured by Russell's paradoxical question, "Is the set of all sets a subset of itself?" The deconstructionist boldly answers "*Il n'y a pas de hors-texte.*" Gross's scheme is less frank but equally susceptible to Russell's challenge. Is the proposal to view science *sub specie rhetoricae* itself merely rhetorical or are we to extend to rhetoric the epistemic privilege we deny to science? By treating science proportionally, as both representation and situated practice, we avoid making a futile distinction between the dancer and the dance.

REFERENCES

Barnes, B., and D. Bloor. 1982. "Relativism, Rationalism and the Sociology of Knowledge." In M. Hollis and S. Lukes, eds., *Rationality and Relativism.*

Cambridge, Mass.: MIT Press, pp. 21–47.

Bridgman, P. W. [1927] 1961. *The Logic of Modern Physics.* New York: Mac-Millan.

Burke, K. 1945. *A Grammar of Motives.* New York: Prentice Hall.

———. 1961. *The Rhetoric of Religion: Studies in Logology.* Boston: Beacon Press.

Coleridge, S. T. 1907. *Biographia Literaria.* Edited by J. Shawcross. Oxford: Oxford University Press.

Dewey, J. 1938. *Logic: The Theory of Inquiry.* New York: Holt & Company.

Doppelt, G. 1983. "Relativism and Recent Pragmatic Conceptions of Scientific Rationality." In N. Rescher, ed., *Scientific Explanation and Understanding: Essays on Reasoning and Rationalists in Science.* London: University of Pittsburgh Press, pp. 107–42.

Einstein, A. 1917. "Kosmologische Betrachtungen zur allgemeinen Relativitäts-theorie." *Preuss. Ak. Wiss. sitzungsber:* 142–52.

Einstein, A., and W. de Sitter. 1932. "On the Relation Between the Expansion and the Mean Density of the Universe." *Proceedings of the National Academy of Science 18:* 213–14.

Forman, P. 1971. "Weimar Culture, Causality and Quantum Theory, 1918–1927; Adaptation by German Physicists and Mathematicians to an Hostile Intellectual Environment." *Historical Studies in Physical Sciences 3:* 1–115.

Gross, A. G. 1990. *The Rhetoric of Science.* Cambridge, Mass.: Harvard University Press.

Latour, B., and S. Woolgar. 1986. *Laboratory Life: The Construction of Scientific Facts.* Princeton: Princeton University Press.

Laudan, L. 1990. "Normative Naturalism." *Philosophy of Science* 57: 44–59.

McGuire, J. E., and T. Melia. 1989. "Some Cautionary Strictures on the Writing of the Rhetoric of Science." *Rhetorica* 7: 87–99.

———. 1991. "The Rhetoric of the Radical Rhetoric of Science." *Rhetorica* 9: 301–16.

Melia, T. 1989. "Scienticism and Dramatism: Some Quasi-Mathematical Motifs in the Work of Kenneth Burke." In H. W. Simons and T. Melia, eds., *The Legacy of Kenneth Burke.* Wisconsin: University of Wisconsin Press, pp. 55–73.

Rescher, N. 1977. *Methodological Pragmaticism,* Oxford: Blackwell.

Rorty, R. 1979. *Philosophy and the Mirror of Nature.* Princeton: Princeton University Press.

Russell, B., and A. N. Whitehead. 1925. *Principia Mathematica.* 1st ed. Cambridge, England: Cambridge University Press.

Stump, D. 1991. "Fallibilism, Naturalism and the Traditional Requirements for Knowledge." *Studies in the History and Philosophy of Science* 22: 451–69.

Worrall, J. 1988. "The Value of a Fixed Methodology." *British Journal for the History of Science* 39: 263–75.

———. 1989. "Fix It and Be Damned: A Reply to Laudan." *British Journal for the History of Science* 40: 376–88.

5

The Strong Program in the Rhetoric of Science

Steve Fuller

Department of Sociology and Social Policy, University of Durham, UK

How does the rhetoric of science look from the standpoint of Ockham's Razor? Self-styled rhetoricians of science have been so good at advertising the field's affinities with literary criticism, discourse analysis, hermeneutics, cultural studies, and relativist philosophies of science that serious scholars of the rhetorical tradition have questioned whether the label "rhetoric" adds anything new (Gaonkar 1993). One reason for proposing a "Strong Program in the Rhetoric of Science" is to alleviate these doubts (Bloor 1976, chap. 1, is the "paradigm" for proposing Strong Programs in science studies).

However, my primary intent is not separatist. The main reason for promoting a distinct rhetoric of science is to challenge the unexamined assumptions increasingly made by the *other* fields in science studies—history, philosophy, sociology, and literary studies of science—especially as these fields have turned to case studies of "science in action" (Latour 1987) as the preferred units of analysis. The case study—a marvel of rhetorical construction—is typically presented in an empiricist spirit of letting the object of inquiry dictate the required theoretical resources. In this way, theory is "grounded." Thus, when conducting a case study, a bricoleur's attitude is commonly taken toward theory such that positions are adopted and discarded as they suit the case. As a result, substantial theoretical differences rarely receive a full airing, but rather are suppressed in a kind of interdisciplinary politesse that travels under the euphemism of "alternative interpretive frameworks" (Fuller 1993a, chap. 2).

A lingering fear and loathing of logical positivism's "unity of science" program has undoubtedly hampered science-studies practitioners from confronting and resolving their theoretical differences. Yet, positivism's top-down metatheoretic management of unruly discourses is not the only possible model here. From its Sophistic beginnings, rhetoricians have been adept at providing the counterpoint, complement, or missing term needed to bring latent disagreements into focus so as to raise the disputants to a new level of discussion. In a similar spirit, I will sharply contrast "the rhetorical standpoint" with the perspectives taken by philosophical, literary, historical, and sociological studies of science. My approach to rhetoric resembles the Sophists (in whose number I include Socrates) in its suspicion of discipline-based understandings of phenomena as polymorphous and public as those associated with science. At the end of this essay, I offer two general questions suggested by this Sophistic orientation, which could launch a Strong Program in the rhetoric of science.

Philosophy

Whenever rhetoricians venture upon theoretical turf, they invariably stumble upon some philosophers, who unfortunately continue to set the terms for their subsequent encounters. Thus, the moves in the ensuing debates tend to fall back into the grooves of philosophical disputes between relativism and realism (or rationalism) that the vaunted case-study method was supposed to have rendered irrelevant. Among rhetoricians, Gross (1990) seems to have been most willing to replay the endgame of those disputes.

However, the cause of rhetoric is better served, not by trying to resolve the old philosophical debates, but by examining the role of these debates in motivating scientific inquiry. Many scientists claim that their work is licensed by a well-grounded commitment to one or another philosophical position. For example, a diminished commitment to realism has been taken to warrant "softer" scientific methodologies.[1] Moreover, the specter of the opposite position—be it of relativist decadence or realist dogmatism—has often instilled an attitude of perseverance said to be psychologically necessary for a line of inquiry to yield all its fruits. However, rhetorical interest in the old philosophical debates need not extend to participating in them. Indeed, the rhetorician may be required to point out that the recent rounds of these debates have introduced a level of sophistication that has failed to register with the scientists who were originally convinced of the importance of declaring for realism, rationalism, or relativism.

Let us address why rhetoricians should not even commit to the relativist position, one with which rhetoricians since Protagoras have been routinely associated. The problem with relativism—at least as it is formulated in recent philosophical discourse—is that it presumes a rather unrhetorical conception of language in which worlds are projected wholesale from so-called linguistic rules. The Sapir-Whorf Hypothesis of linguistic relativity is a case in point. It is modeled on the different spaces projected by alternative axioms of geometry. In the hands of the founder of structural linguistics, F. de Saussure, this model was used to reinforce the old philological practice of reconstructing the grammars of dead languages as clues to past *Weltanschauungen*.[2] However well these models may have served geometry and philosophy, they serve the rhetorician poorly. Wittgenstein realized as much as he gradually changed his views on the nature of language over the course of his philosophical career.

The philosophical relativist is typically portrayed as claiming that truth and goodness are determined by the rules, or "norms," of the society in which one is located—as if these norms applied the same way in all cases, and hence could be used as criteria to demarcate one society from the next. Instead of embracing this position, the rhetorician should inquire into the situations that call for explicit statements of norms in the first place, and for whose benefit are such statements made. If the law and anthropology are good guides, normative statements would seem to be made for the benefit of (a) visiting strangers, (b) new members, and/or (c) posterity. Conspicuously absent from this list are active members in the society. Their governance is a subtler matter, one in which normative discourse plays only a peripheral role until some significant violation is recognized (Fuller [1989] 1993, chap. 4). Thus, the rhetorician should not expect that native behavior could be predicted simply on the basis of knowing canonical statements of native norms. Of course, the reasons such normative statements are made, despite their failure to represent actual practice, are not trivial, as they are bound up in complex issues of the collective self-image that a society wishes to project. However, these are not issues that typically enter into philosophical disputes over relativism.

As we move from philosophy of language to philosophy of science, despite protests to the contrary, most of its analytical practitioners continue to think in the shadow of the positivist image of scientific language that is profoundly unrhetorical, yet familiar from our previous discussion. The image is one of language as consisting of a mapping function from one syntax to another (i.e., its semantics), which is then embedded in a speech context, which, in turn, determines how the language is used in particular situations. Although few philosophers now take the idea of pure syntax seriously, they still adhere to the pecking order implicit in the old image. Thus, nowadays science is envisaged to proceed by constructing a theory (syntax) which projects an ideal domain of objects (semantics), whose application to empirical reality is then negotiated in experimental practice (pragmatics).[3] Philosophers in search of a place for rhetoric in this image of scientific language have typically identified rhetoric with the final, pragmatic stage.

A case in point is P. Kitcher's (1991) account of the role of rhetoric in a geometric proof. According to Kitcher, rhetoric's role lies in explaining the difference in the way one would present a proof to enable a novice and an expert to understand its line of reasoning. Since the student is unfamiliar with the canonical formulation of the proof, steps must be provided that the expert would regard as logically trivial. Rhetoric, so says Kitcher, involves fine-tuning the expression of an epistemic entity to a target audience. The epistemic entity—in this case, the proof—is typically established by nonrhetorical means, such as deductive logic as interpreted through the semantics of, say, Euclidean geometry.

Within the philosophy of mathematics stands a position that does a better job than Kitcher's in approximating a robustly rhetorical stance. This position, associated with the school of "intuitionists" and especially "strict finitists," makes the intelligibility of a chain of reasoning to a "natural consciousness" a necessary part of what makes that reasoning a proof. This "natural consciousness" functions much as the legal fiction of the "reasonable man" [sic] in Anglo-American jurisprudence—someone who is not a complete mathematical illiterate but who has retained enough common sense not to be overly impressed by the ex cathedra hand-waving and gratuitous technicalities of expert mathematicians. Such a sensibility is perhaps most familiar to readers of the later Wittgenstein's *Remarks on the Foundations of Mathematics* (see also Bloor 1983).

The rhetorical character of this alternative account of mathematical reasoning is threefold. First, the status of an epistemic entity, the proof, depends on its acceptability to an idealized audience, the natural consciousness. (On some readings of Wittgenstein and strict finitism, the audience would be the actual witnesses to the reasoning, but I will presume an idealized audience.) Second, this hypothetical audience is designed to counteract the tendency of a relatively autonomous epistemic community—that of professional mathematicians—to produce knowledge that will have purchase only for its own members. Just as the legal fiction of the reasonable man was designed to remind jurists of whom the law is for, so too the postulation of a natural consciousness is supposed to remind mathematicians that their constituency extends considerably beyond their own expert practitioners.[4] Indeed, intuitionists and strict finitists have gone to great lengths to insure that mathematics remains a genuinely public practice—even to the extent of eliminating transcendental numbers and those parts of set theory for which no constructive proofs can be given. Finally, the account presumes that the composition of a natural consciousness changes over time, or at least one can show that its composition has changed. This point results from the fact that the natural consciousness is nothing more than what it does in particular cases. No a priori structure to the natural consciousness is presumed, only the record of its actual decisions. (In my characterization of mathematical intuitionism, I exclude L. Brouwer's original Kantian formulation of the natural consciousness, which implied a priori psychological limits to mathematical reasoning.)

Literary Studies

It may seem odd that I would want to distance the rhetoric of science from literary analyses of texts, since much of the pioneering work in rhetoric of science has been done by people with strong literary backgrounds (e.g., Bazerman 1988, Gross 1990) who treat scientific writing as a distinct genre, within which one can identify tropes, figures, and devices that characteristically frame scientific arguments. While, of course, we have considerable historical precedent—going back to Aristotle—for identifying rhetoric with textual interpretation, making a sharp distinction between a rhetorical and a literary approach to science is useful for the "Strong Program". The reason is informed by an analogy from psychology, a discipline whose history has been dominated by what J. Fodor (1981) has called "methodological solipsism," namely, the idea that the psychological subject can be exhaustively studied by confining one's inquiries to the subject's mental universe. Analogously, the literary approach encourages a "textualist" approach to rhetoric that is just as artificial as the psychologist's solipsism.

While solipsism gave psychology a clearly defined domain of inquiry, it did so by making it the natural heir of the rationalist-empiricist debate in philosophy, which had rested on some dubious assumptions including the idea that the mental and nonmental are radically different in kind (Rorty 1979). Thus, in the modern period, the relative autonomy of psychology from the other natural sciences has been bought at the price of rendering the language of mental life largely incommensurable with the language of physical reality. Moreover, both philosophers and psychologists routinely assume that this incommensurability is not merely verbal but reflects a deep metaphysical problem—traceable to Descartes—of how the subject gets outside its mind to apprehend a nonmental reality. However, in light of psychology's disciplinary history, I concur with J. Gibson's "ecological" school that this is probably a pseudoproblem in the sense of not reflecting the ontological situation of organisms in the world. Rather, "the problem of the external world" merely reifies the mutually exclusive disciplinary paths that psychology and the natural sciences have taken since the seventeenth century. In that case, even the basic idea that organisms transform—by special "cognitive" means—raw physical parameters into psychologically meaningful units may turn out to be little more than a sophisticated act of ventriloquism if the place in the organism reserved for its cognitive powers merely inscribes the translation problem facing psychologists and natural scientists.

To appreciate the analogy between textualism and methodological solipsism, we need to return to the original attempt by the Russian and Prague Formalists to establish a "science of the literary" (Galan 1984). This science purported to treat the text as a closed universe of discourse where each syntactic distinction was taken to make a semantic difference. Methodologically speaking, this meant that the literary scientist was to read all texts with the same closeness as one would read poems: Each word was presumed to serve a distinct semantic function in the poem, with even apparent instances of redundancy taken to do more than simply reiterate an earlier point. In its day, in the first two decades of this century, the Formalists' textual imperialism was revolutionary in its suggestion that any text could be productively read with the same care and intensity normally reserved for the products of "high culture." Indeed, this thesis was crucial to the historical point Formalists wanted to make about the way exemplary authors must have read popular literature in order to translate its forms into texts intended for more elite audiences. Thus, high culture does not constitute a self-sustaining tradition but continually draws upon the so-called lower genres for revitalization.

Note that just as the commitment to methodological solipsism forced psychologists to internalize obliquely—by ascribing "cognitive powers" to the organism—the features of reality that had been originally excluded from their domain, so too the commitment to methodological textualism has caused literary critics, in the twentieth century, to simulate qualities and processes that would normally be thought to lie beyond the text proper. Thus, the psychic quality of "genius" and the social process of literary transmission are simulated in reading practices that approximate those of the disciplined Formalist critic. This is not to deny the interest of the studies that have been done under textualist constraints, but to advise that their warped historical vision may do a disservice to the rhetorician of science.

Whatever may be true of the reading habits of literary cultures, we have little evidence that scientific texts are actually read with the sort of care and intensity to which the textualist aspires. It may be little more than a humanist conceit to suppose that the reading practices humanists value the most would be shared by the most highly esteemed knowledge producers in our society, the natural scientists. After all, while texts are the primary objects of inquiry in the human sciences, they are, at most, means to other ends in the natural sciences. Indeed, a lasting legacy of the Scientific Revolution was to instill in scientists a professional impatience toward argument and the other verbal arts as so much time taken from the construction of instruments needed to demonstrate findings that would hopefully *silence* opponents (Shapin and Schaffer 1985). Unsurprisingly, the history of scientific journal writing exhibits successive innovations to streamline the reading process. These aim at modularizing textual presentation—well-defined sections on "theory," "method," "data," "discussion"—so as to enable scientific readers to appropriate relevant sections for their own purposes, and to ignore (not criticize) the rest (Bazerman

1988). Indeed, what would strike the humanist as a negligent textual practice—namely, to read only the highly cited section of an article—appears to be crucial to the consolidation and cumulative growth that distinguish knowledge in the natural sciences from knowledge in, say, the social sciences. In the latter, a highly cited article would be read more closely, but also with less agreement over what makes it so noteworthy (Cozzens 1985).

Now, of course, argument is sometimes not a diversion in science, but the very center of scientific activity, especially when a paradigm is in "crisis." However, we would be mistaken to take the dominance of the verbal arts in these periods to imply that scientists suddenly acquire humanists' scruples in what they read and write. To make undue assumptions about scrupulousness is to invite spurious historical questions about the scientists' rhetorical situation.

Consider a frequent question about the reception of Darwin's *Origin* (1859): Why did Darwin's critics fail to appreciate the cogency of his arguments for natural selection? The answers typically given turn on claims about the biases or incapacities of Darwin's readers, as well as corresponding claims about Darwin's own depth and foresight (or perhaps his text's "unparalleled richness"). Yet, the appropriateness of this question rests on several simple but historically unwarranted assumptions about Darwin's rhetorical situation: (1) that Darwin's critics were obliged to read *Origin* as closely as the humanist does now; (2) that the primary cognitive interest of Darwin's critics was to evaluate Darwin's claims; (3) that the criticisms made of Darwin should be taken as primarily addressed to Darwin and his followers.

These three assumptions underwrite a "book club" model of scientific rhetoric, as if one should think of *Origin*'s reception as an event focused on the text itself. *Origin* would better be thought of as having been thrown into the middle of many ongoing debates, subject to the vicissitudes of several parties trying to get whatever mileage they can out of what the book says. In that case, the rhetorically interesting feature of *Origin* lies in its ability to restructure the debates in which it so variously figured; for even as Darwin's opponents contested the specific doctrine of natural selection, they typically adopted enough of *Origin*'s language to presume that some version of the general doctrine of evolution is true (Ellegard 1990). Thus, in the 30 years following the first edition of *Origin,* one can witness a subtle but real shift in the burden of proof from evolutionists to creationists. This process can be explained in terms of the associations that disputants were able to make between *Origin*'s language and their own concerns. And such explanations can be offered without the rhetorician of science having to impute a spurious psychology to either Darwin or his readers (Fuller 1993b, responding to Campbell 1987).

Upon encountering the brute character of scientific reading and writing practices, the die-hard textualist has three options. The first, common among rhetoricians of science of a literary bent, involves tacitly replacing the real readers with an ideal reader, someone knowledgeable of the science under discussion but who is also blessed with a humanist's sense of interpretive care. As we have seen, this option invites the specious forms of historical understanding noted earlier. However, a second option available to the textualist is to admit that everyone routinely misreads each other's texts. Ever since deconstructionists redefined poetic and philosophical originality as the ability to provide "strong misreadings" of one's distinguished predecessors (Bloom 1972), this view has held a certain attraction. At a more mundane level, historians interested in reconstructing a debate from original documents often unwittingly stumble into this option, which, in their telling, may reduce a Manichaean struggle to a comedy of errors: The great minds would have realized they were not so far apart had they bothered to read each other's works a little more closely. At this point, how we can claim to have made epistemic progress is a complete mystery. (Despite its surface outlandishness, option two usefully shows what happens when certain taken-for-granted notions of communication are put to a serious empirical test [Fuller 1988, part 2]. However, analytical philosophers, in the spirit of Quine and Davidson, are likely to balk at this suggestion, arguing that a theory of communication according to which everyone miscommunicates simply means that communication is not adequately modeled by the theory— not that communication itself fails to occur.)

Enter the third option, which promises moderate relief from this epistemic predicament. It openly accepts that textualism is nothing more than a normative orientation toward scientific reading and writing practices. The claim would be, then, that science would be a better enterprise *if* scientists were to adopt the textual practices of humanists, but scientists would not be presumed to naturally do so. Such a claim could be grounded in a social theory—for example, Habermas's theory of communicative action—that takes the quality of language use to be an index of the quality of the human condition. One potential ally in this project would be P. Feyerabend, who argued repeatedly that the political and economical stakes tied to contemporary "Big Science" prevent scientists from treating each other's arguments with the care they deserve: Because too much money—and too many people's careers—ride on my being right, I am compelled not to take seriously the possibility that I may be wrong.

As noted earlier, Formalism's rigorously textualist approach was radical in the links it revealed between popular and elite literary forms. Unfortunately, a text-based rhetoric of science would only obscure such links, which nevertheless do exist, because the situation is now reversed. Instead of showing how popular forms can be productively read in the manner of elite forms, the rhetorician faces that the production and consumption cycles of texts in science have come to resemble those observed in the mass media. Indeed, studies of the *Science Citation Index* reveal that the "harder" the science, the more its research specialties resemble fads in their life cycles. By contrast, research specialties in the "softer" sciences have long half-lives, so that when they definitively go out of fashion is often unclear. This finding is unsurprising given the similarities in the size and shape of today's scientific enterprise and of the mass media. One witnesses both "invisible colleges" and "opinion leaders" which structure the reception and appropriation of texts in both domains (De Mey 1982, chaps. 7–9). Whether this should cause the rhetorician to devalorize science or valorize mass media—or reassess our general value orientation toward different forms of knowledge—is an open question, but no less important to pose.

One methodological precaution can be taken by rhetoricians of science to avoid the traps of textualism. It involves comparative studies of reception: Why was one text used in ways that significantly restructured debate in some domain of inquiry, whereas another text—ostensibly vying for the same span of collective attention—failed to do so, or had some other effect instead? For example, Fuller (1993a, chap. 4) attempts to explain why texts by A. Marshall rather than by G. Wallas succeeded in laying a foundation for a "science of politics" in Britain, even though the former was an economist and the latter a political practitioner and theorist. The weakness of the literary approach to rhetoric would become evident in such comparative studies as the rigorous textualist should be able to show that, read with sufficient care, *either* text *could* have had the desired impact on its audience. So why, then, did only *one*? (Another relevant comparison is to examine the fate of the same text in two different fields of argument, such as the reception of Keynes's *General Theory* [1935] among academic economists and among policy makers. Diesing 1991, chap. 5, and Stehr 1992, chap. 4, attempt such an analysis.)

History

Although H. White's (1972) *Metahistory* has been subject to a wide range of critical responses, it succeeded in making "rhetoric" a respectable topic of interest for professional historians. However, White opened the door to a certain literary conception of rhetoric whereby the historian was said to adapt her understanding of the past to one of a fixed number of universal plot structures. This focus on "emplotment" has encouraged historians to stress narrative time to the near exclusion of real time—what a classical rhetorician might be inclined to call a very high *mythos*-to-*kairos* ratio (Kinneavy 1986). Another way of making this point is to observe that a higher premium is placed on the internal logic of events recounted in the historical narrative than on why the sequence transpired at the pace and over the length of real time that it did. Moreover, self-styled "contextualist" histories, ones heavy in Geertzian "thick descriptions," tend to reduce temporality to overlapping cultural tendencies in which even the actual order of events no longer seems important (Bender and Wellbery 1991). What is in danger of being lost here is the rhetorician's concern for the *timing* of events: why something occurs at the specific time that it does.

Of course, some histories of science have been sensitive to timing, most notably P. Forman's (1971) account of the adoption of the indeterminacy interpretation of quantum phenomena by physicists in Weimar Germany. Unsurprisingly, given the general academic insensitivity to timing, Forman's thesis has often been misinterpreted. He is often taken to have argued simply that the doctrine of quantum indeterminancy was a "reflection" of the irrationalist tendencies in Weimar culture. The obvious counterargument is that ideas of quantum indeterminancy predate Weimar culture, thereby establishing the relative detachment of those ideas from any specific social setting. However, Forman actually acknowledges this point at the outset of his inquiry. However, he is less concerned with underlying causes than with occasioned action: Granted that the crucial ideas had been debated inconclusively for at least 50 years, why do physicists, all of a sudden, close ranks behind the indeterminacy interpretation in 1927? Forman purports to show that no new arguments or experimental findings justified the quick closure of debate. Instead, we need to look at the growing cultural hostility to determinism and materialism (both associated with the German loss in World War I), which was about to threaten the funding of physics research. While we may question whether physicists were as calculating as Forman describes, Forman nevertheless deserves credit for asking the sort of historical question that would be expected of a rhetorician of science.

Let us consider how alien Forman's mode of inquiry is to the way historians and philosophers of science have traditionally framed the problem of theory choice. But first a disclaimer: Social historians of science have demonstrated the artificiality of the idea of a "problem of theory choice" insofar as it assumes two or more well-defined groups of scientists lined up behind fully articulated research programs. History is rarely that neat and clean. However, setting that issue aside, one can certainly detect many long-standing theoretical disputes in the history of science. Yet, what is at stake is often unclear, aside from a point about the right way to talk about a certain class of phenomena. Not surprisingly, then, these disputes can simmer for long periods, causing little disruption to the business of doing experiments and collecting data. One might be tempted to follow Hacking (1983) in claiming that the relative autonomy of theoretical disputes demonstrates their relative *lack* of influence on the course of scientific research.

Nevertheless, the dispute eventually reaches a climax—the paradigm is in crisis, a revolution ensues, and closure is finally reached. Traditionally, historians and philosophers have presumed that the time for closure is always appropriate, and that the interesting question to ask is what were the decisive arguments and evidence that favored one side over the other(s). By contrast, a rhetorically sensitive historian like Forman would ask why the "moment of decision" occurs at this point in time and not some other. After all, had the moment been a bit earlier or a bit later, the relative standing of the parties to the dispute—not to mention the actual composition of the parties—may have been different. A slight change in context, and the weaker argument may have been, indeed, the stronger.

If we imagine that all of the history of science could be written in

Forman's rhetorically sensitive manner, then a subversive hypothesis is suggested: Those defining moments of scientific culture—the great paradigmatic showdowns—occur only because science does *not* always follow its own internal trajectory. To borrow a current biological image, external pressures provide the "punctuation" needed to establish new internal equilibria for science. In a sense, this point is implicitly conceded by the infamous Duhem-Quine thesis. Left in the realm of dialectic, for any two competing theories, every new argument or piece of evidence can be met by a counter or equivalent piece. But fortunately, every so often, these matters of dialectic get caught up with external matters of action, the stuff of rhetoric, which force the parties to settle their differences summarily. In short, she who controls *when* the decision is made controls *what* decision is made. (Levine 1989 offers social psychological evidence for this claim.)

As Forman's reception suggests, closely related to the failure to distinguish clearly narrative time (*mythos*) from real time (*kairos*) is a fundamental ambiguity about how "the social" should be conceptualized in historical explanation. On the one hand, much of the so-called social history of science is really a history of scientific traditions. In terms of its units of analysis and frames of reference, such history is not unfairly seen as a fleshed-out version of the old internal history of science (e.g., Galison 1987). Its philosophical sustenance is drawn largely from Polanyi and Kuhn, and its methodological forebears are more easily seen in *iconography* than in any other branch of history. The key to this approach is that episodes and artifacts in the history of a given science are explained as contextualized reproductions and transformations of prior episodes and artifacts in the same history. Thus, history appears to offer new twists or combinations to time-honored themes.

On the other hand, opposed to this *mythos*-based approach are social histories of science that stress the exigent character of what scientists do, namely, their need to construct opportune responses to situations that are largely *not* of their own creation. "The tradition" no longer appears as a continuous medium of cultural transmission, but as a stockpile of contestable resources. In Forman, as well as in Shapin and Schaffer (1985), science is described as the art of using intellectual means to solve social problems. Here scientists always need to repair the integrity of their practices by reframing outside pressures as problems that can be legitimately treated within their own domain. Although a strongly rhetorical orientation would dictate that historians prefer this latter, *kairos*-based approach, the tendency toward mythic history has served both traditional scientific and humanistic interests in writing the history of science.

Consider the traditional scientific interest first. The professional identity of modern science was forged largely by accepting certain forms of mythic history as its own. Proctor (1991) recounts this process, which typically occurred as part of a forced exchange whereby scientists agreed to ignore certain sensitive features in their environment in return for political protection. The Charter of the Royal Society of Great Britain is the archetype of all such agreements as it compelled the first organized scientific community to relinquish any investigative interest in politics, morals, religion, rhetoric—in short, anything that might impinge on the legitimacy of established authorities. More recently, similar agreements have been announced in the name of "value neutrality" and "academic freedom." Of course, none of these forced exchanges diminished the interest of scientists in making timely interventions into public affairs. However, in order to maintain their professional ethos, scientists have had to try to colonize public problems as extensions of their particular fields, which, in practice, meant recounting such problems as emerging naturally from their disciplinary narratives. The success of this practice has typically benefited scientists more than the public.

Humanistic interest in promoting a *mythos*-based history of science is similar in content but of more recent vintage. Iconographic approaches take hold in various branches of the humanities when scholars interpret the irrationalism surrounding the outbreak of World War I as a frontal assault on the cultural heritage of Europe (Hart 1993). These approaches are especially ascendant in Central Europe, with many of the leading representatives migrating to England and America with the outbreak of World War II. They include E. Panofsky (art history), K. Mannheim (sociology of knowledge), L. Spitzer (literary history), L. Strauss (political philosophy), and M. Polanyi (philosophy of science). All of these thinkers share an interest in showing how cultural artifacts manage to survive the troubled circumstances of their creation in order to become part of our common heritage. Though considerable differences of nuance exist between them, they all write as if these artifacts were "doubly encoded." The deeper code is typically one that would subvert conventional life were it fully revealed, but it happens to be overladen with a surface code that appeals to, or at least positively impresses, conventional tastes. To understand the deeper code, one needs to be thoroughly immersed in the traditions upon which the artist or scientist draws.

But what is so subversive about these traditions that forces scientists and artists to act so cultlike? The general answer is a Platonic concern for the peculiar position of elites in a democratic culture. Elites must publicly legitimize their activities in terms that are, at best, orthogonal, and perhaps even contradictory, to their self-appointed missions. For example, democracies lavishly fund scientists because science is persuasively portrayed as the pinnacle of public service. But scientists are called to their vocation by a commitment to fathoming mysteries that, by their nature, must remain remote from the concerns of ordinary mortals. For Polanyi (1957), the continuity of science is maintained, on the one hand, by the respect (if not awe) that lay people have for the technical wizardry of science, and, on the other hand, by the care with which scientists monitor each other's actions so that the relevant traditions are transmitted without corruption. The threat of corruption is ever present because scientists are always tempted to believe their public image, succumb to hubris, and invariably suffer a tragic fate. Thus, in World War I, the German scientific community erred in thinking that intellectual superiority conferred some natural advantage in combat. The German public all too willingly indulged these delusions and then all too willingly blamed the scientific community once Germany lost the war.

Kuhn interestingly complicates the humanist's tragic vision of cultural transmission because of his commitment to periodic revolutionary upheavals in the history of science. Yet, Kuhn [1962] 1970, 167) stresses that these upheavals must be as short and contained as possible, and that new generations of scientists be taught an "Orwellian" (Kuhn's word) history of the revolution that rewrites the period fully within the narrative canonically told about the discipline's development. In fact, in his historiographic works (e.g., Kuhn 1977), Kuhn tends toward the view that the kind of rhetorically sensitive history associated with Forman is incompatible with the pursuit of normal science because the latter receives direction and meaning from the kind of Orwellian history which the former aims to deconstruct. For this reason, despite his graduate training in physics, Kuhn has been unable to write histories of the period after World War I since he continues to participate in the mythic vision of the discipline that began to form at that time (see Kuhn 1978). Indeed, partly in order to distance himself from his admirers in the humanities and social sciences, Kuhn has made a point of arguing the impossibility of doing good history of a period in which one has a stake in the outcome (Fuller 1992). This antikairotic sentiment runs very much against the Strong Program in the rhetoric of science.

Sociology

As the story is normally told, rhetoric was a direct beneficiary of the inroads that sociologists made into philosophical accounts of science. Rhetoricians were issued an open invitation to study science when the normative conceptions of Truth, Rationality, and Objectivity metamorphosed into truths, rationalities, and objectivities; when it was conceded that science shared many of the characteristics of other social practices; and especially when science was shown to be a concept whose definition is routinely contested by various parties, scientists being only one of them. While "the sociological turn" (Brown 1984) has provided a necessary background condition for the emergence of the rhetoric of science, it is not sufficient for sustaining our Strong Program.

What needs to be made up is sociology's failure to deliver on its nineteenth-century promise of a form of knowledge that would enable social transformation as it provided a systematic understanding of the modern world—a *knowing that* that was simultaneously a *knowing how*. This aim was expressly sought not only by Comte and Marx, but even by such professional academics as the Durkheim School in France and the founders of American sociology departments (Ross 1991). The promised knowledge would have been strongly rhetorical. Unfortunately, much of academic social science today—not just sociology— remains rhetorically inert by enlightening audiences in ways that do not appreciably enhance their abilities to alter their social situations (Fay 1987).

The rhetorical incapacity of the social sciences bears the mark of Aristotle's original strictures on the domain of rhetoric. Consider the simple idea that knowledge must be grounded in the nature of the object of knowledge, which is taken to exist prior to the communicative act. This seemingly innocuous notion presupposes that a clear distinction can be drawn between knowledge proper and the (rhetorical) means by which it is communicated, which, in turn, suggests that the object of knowledge remains fundamentally unchanged by the communicative process. On this view, the strength or weakness of a knowledge claim is relative to a standard presumably set by the object of knowledge and not by the audience to whom the claim is addressed. It thus becomes difficult to see how knowledge by itself could motivate action.

Not surprisingly, then, most social scientists presume the possibility for an explanation of the social order to be adequate to the facts yet fail to offer any guidance as to how that order may be changed. Indeed, it may especially fail to provide guidance to those in whose interest it would be to change that order, namely, the oppressed groups in the society. This divorce of knowledge from action occurs at two levels: at the level of "lawlike regularities" that apply to units of social analysis too large or too small to be within the control of the people subsumed under the regularities, and at the level of esoteric technical discourse, which places the regularities beyond even the intelligibility of the ordinary person.

One supposed advantage of accepting this incapacitating conception of the object of social inquiry is the predictability of social life that it entails, which can be turned into an administrative tool for "normalizing" individuals. Of course, in practice, this technocratic vision has been less than a complete success, but the Aristotelian essentialism that informs the vision remains sufficiently potent to discourage most people from mounting organized resistance to the social order (Popper 1957, Unger 1987). Many of society's doors may be unlocked, but social science discourages most from trying to turn the knobs.

This point applies no less to academics than to the more traditionally disenfranchised elements in society. The problem is that academics are enjoined not to intervene in ongoing social processes until they have acquired a comprehensive understanding of how those processes work, by which time the academic is often persuaded that the independent character of the processes transcends her ability to influence them. For example, Marxists who pursue "historical materialism" as a research program that is propaedeutic to any political agenda typically become convinced that one must *wait*, rather than act, for the revolutionary moment to occur. The fear of error from acting on a partial understanding inhibits the kairotic impulse to take calculated risks and seize opportunities. The prospect of subsequently having to adapt to circumstances is seen as a symptom of prior cognitive deficiencies rather than as an occasion to construct new courses of action. All of these tendencies add up to a vision of social inquiry as endlessly temporizing and self-inhibiting.

In the case of the recent social studies of science, this vision is often associated with a Wittgensteinian attitude of social inquiry "leaving the world alone." One of the field's leading practitioners, H. Collins (1987), has argued that the success of sociologists in demystifying philosophical rhetoric about science should not be taken in the spirit of delegitimizing scientific authority; rather, it should be seen simply as an empirically more adequate account of science that may unburden the public of certain unrealistic expectations about what science can deliver. But such accounts, Collins maintains, should leave the future of science largely in the hands of the practicing scientists. Collins sidesteps the very question that someone who believed in the constitutive role of rhetoric in the scientific identity would ask: Given that scientific authority has been so closely tied to its philosophical rhetoric, why should its demystification *not* be taken as an opportunity to renegotiate science's social contract (Fuller 1994)?

Two General Questions for the Strong Program

In conclusion, I offer two general questions to pull together the distinctive strands of the Strong Program in the rhetoric of science into a unique research agenda.

1. *Does the culture of contestation trivialize knowledge claims? And if so, what does that tell us about the nature of scientific knowledge?*

Virtually every philosopher who has claimed to speak on behalf of "scientific rationality" has endorsed some version of the idea that criticism is the lifeblood of epistemic growth. Yet, philosophers (not to mention scientists) typically recoil whenever it has been suggested— by, say, Feyerabend and sometimes Popper—that any knowledge claim should be open to criticism at any time. The fear is that inchoate or controversial research programs may be prematurely terminated for their failure to answer criticisms at the time they are lodged. And even if the criticisms miss their target, they may leave enough doubts to hinder the target program from development. Moreover, if the activity of lodging criticisms is too highly valued, then inquirers will lose all incentive to develop their own knowledge claims beyond a certain point. For these reasons, philosophers have tried to derive ways of protecting certain kinds of knowledge claims from immediate attack. Sometimes they appeal to logic, sometimes to history, and increasingly to expertise, but the net effect is to defer the critic's access to the forum.

The rhetorical intuitions informing this philosophical sensibility are worth probing. First, criticism would seem to be denied a creative role in thought: At most, it is a selection mechanism. Philosophers prefer theories that are reasonably developed before being subjected to criticism, while they regard theories that receive most of their elaboration in the face of criticism as having been ill-conceived or "ad hoc"—despite historical cases to the contrary.[5] A rhetorician would start to suspect that a big part of the mystique of science lies in its dialectical encounters being portrayed as do-or-die struggles that occur relatively rarely because of the magnitude of the stakes, which is reflected in the degree to which a theory must be developed before being brought to the forum. Conversely, respect for science could conceivably diminish if too many of its knowledge claims were too frequently contested.

2. *Is it rhetorically possible to write a book today that would have an impact on the conduct of scientific inquiry comparable to that of Newton's* Principia *or Darwin's* Origin? *And if not, what does that tell us about the nature of revolutionary change in science?*

To cast this question as one of *rhetoric* is to challenge certain tacit assumptions common to most science-studies practitioners. Two reasons are routinely offered for why, in recent times, no books have been as influential as Newton's or Darwin's. One reason is simply that geniuses of the caliber of Newton and Darwin are rare; the other is that the relevant set of sciences have not yet developed sufficiently to enable the possibility of a Newton- or Darwin-like synthesis. In contrast, the Strong Program in the rhetoric of science would advance a third reason for consideration: No one with the interest, energy, and skill to forge such a synthesis knows how to write in a way that would simultaneously address all the audiences whose cooperation would be necessary for turning that synthesis into a new paradigm for normal science. (Newton, whose formidable mathematical style alienated readers, was able to recruit people to make that style more accessible.)

Newton and Darwin flourished at two ends of a period in European history when access to written materials was sufficiently widespread that scientific authors could presume that readers came to their texts with a prior understanding of certain natural phenomena upon which the authors could then build their cases. This understanding typically resulted from readers having relative access to more elementary works on a topic. Thus, authors did not need to recapitulate the entire domain of natural history before proceeding to their own contribution (Eisenstein 1979). However, the spread of the printed word had not yet reached the point of servicing the highly differentiated and mutually unintelligible communities of readers, as we have now. Part of that history is told by the increased specialization of the scholarly community, but the pace and intensity with which it has occurred—something to which a rhetorician would be sensitive—must take into account the metamorphosis of commercial publishing into the first "post-industrial" enterprise (Horowitz 1986).

NOTES

1. An important exception is Harré and Secord ([1972] 1979), for whom a qualitative approach to the social sciences is grounded in an Aristotelian realist sense of human agency. Indeed, they argue that positivism's own diminished commitment to realism—a preference for correlation over causation—is responsible for its conception of social science as an instrument of prediction and control. See Manicas (1986) for a history that supports this minority viewpoint.

2. Saussure (1857–1913) was trained in both physics and philology, and explicitly saw linguistics as the foundational science of the "social physics" being pursued at that time by Durkheim and other positivist sociologists.

3. Note that this image is presupposed even by the so-called Duhem-Quine Thesis of the underdetermination of theory choice by evidence, which has been used to underwrite most of the epistemological criticisms that sociologists have made of philosophers of science (e.g., Collins 1985). Yet upon close examination, the doctrine makes merely residual space for the "pragmatic" factors so dear to sociologists and rhetoricians—that is, only once the properly "epistemological" or "cognitive" factors have exhausted their potential to explain a theory choice.

4. The history of the law is a good source for thinking about the normative structure of science from a rhetorical standpoint, especially once a necessary condition for the legitimacy of a norm is assumed to be the consent of those who would be governed by it. Fuller (1985) provides an extended discussion of attempts—especially by Napoleon—to make the "natural intelligibility" of the law a precondition for its acceptance, and its possible implications for the governance of science.

5. Reflecting the barrage of criticism he had received, Darwin's *Origin* underwent massive revisions from its first (1859) to its sixth (1872) edition. Only half of the original book remained in the sixth edition. For more on the reception history of this and other scientific classics, see Gjertsen (1984).

REFERENCES

Bazerman, C. 1988. *Shaping Written Knowledge*. Madison: University of Wisconsin Press.

Bender, J. and D. Wellbery, eds. 1991. *Chronotypes: The Construction of Time*. Palo Alto: Stanford University Press.

Bloom, H. 1972. *The Anxiety of Influence*. Oxford: Oxford University Press.

Bloor, D. 1976. *Knowledge and Social Imagery*. London: Routledge & Kegan Paul.

———. 1983. *Wittgenstein: A Social Theory of Knowledge*. Oxford: Blackwell.

Brown, J. R., ed. 1984. *The Rationality Debates: The Sociological Turn*. Dordrecht: Kluwer.

Campbell, J. A. 1987. "Charles Darwin: Rhetorician of Science." In J. Nelson, A. Megill, and D. McCloskey, eds., *The Rhetoric of the Human Science*. Madison: University of Wisconsin Press, pp. 69–86.

Collins, H. 1985. *Changing Order: Replication and Induction in Scientific Practice*. London: Sage Publications.

———. 1987. "Certainty and the Public Understanding of Science." *Social Studies of Science 17*: 687–702.

Cozzens, S. 1985. "Comparing the Sciences: Citation Context Analysis of Papers from Neuropharmacology and the Sociology of Science." *Social Studies of Science 15*: 127–53.

Darwin, C. 1859. *The Origin of Species by Means of Natural Selection*. New York: Hurst & Company.

De Mey, M. 1982. *The Cognitive Paradigm*. Dordrecht: Kluwer.

Diesing, P. 1991. *How Does Social Science Work?* Pittsburgh: Univeresity of Pittsburgh Press.

Eisenstein, E. 1979. *The Printing Press as an Agent of Change*. Cambridge, England: Cambridge University Press.

Ellegard, A. 1990. *Darwin and the General Reader*. Chicago: University of Chicago Press.

Fay, B. 1987. *Critical Social Science*. Ithaca: Cornell University Press.

Fodor, J. 1981. *Representations*. Cambridge, Mass.: MIT Press.

Forman, P. 1971. "Weimar Culture, Causality, and Quantum Theory: 1918–1927." *Historical Studies in the Physical Sciences 3*: 1–115.

Fuller, S. 1985. *Bounded Rationality in Law and Science*. Ph.D. dissertation, University of Pittsburgh.

———. 1988. *Social Epistemology*. Bloomington: Indiana University Press.

―――. 1992. "Being There with Thomas Kuhn: A Parable for Postmodern Times." *History and Theory* 31: 241–75.

―――. [1989] 1993. *Philosophy of Science and Its Discontents.* 2d ed. New York: Guilford Press.

―――. 1993a. *Philosophy, Rhetoric, and the End of Knowledge: The Coming of Science and Technology Studies.* Madison: University of Wisconsin Press.

―――. 1993b. " 'Rhetoric of Science': A Doubly Vexed Expression." *Southern Communication Journal* 58: 306–11.

―――. 1994. "Can Science Studies Be Spoken in a Civil Tongue?" *Social Studies of Science* 24: 143–68.

Galan, F. 1984. *Historic Structures.* Austin: University of Texas Press.

Galison, P. 1987. *How Experiments End.* Chicago: University of Chicago Press.

Gaonkar, D. 1993. "The Idea of Rhetoric in the Rhetoric of Science." *Southern Communication Journal* 58: 255–93.

Gjertsen, D. 1984. *The Classics of Science: A Study of Twelve Enduring Scientific Works:* New York: Lilian Barber Press.

Gross, A. 1990. *The Rhetoric of Science.* Cambridge, Mass.: Harvard University Press.

Hacking, I. 1983. *Representing and Intervening: Introductory Topics in the Philosophy of Natural Science.* Cambridge, England: Cambridge University Press.

Harré, R. and P. Secord. [1972] 1979. *Explanation of Social Behavior.* 2d ed. Oxford: Blackwell.

Hart, J. 1993. "Erwin Panofsky and Karl Mannheim: A Dialogue on Interpretation." *Critical Inquiry* 19: 534–66.

Horowitz, I. L. 1986. *Communicating Ideas.* Oxford: Oxford University Press.

Keynes, J. M. 1935. *General Theory of Employment, Interest, and Money.* New York: Harcourt Brace.

Kinneavy, J. 1986. "*Kairos:* A Neglected Concept in Classical Rhetoric." In J. Moss, ed., *Rhetoric and Practice.* Washington, DC: Catholic University Press, pp. 79–105.

Kitcher, P. 1991. "Persuasion." In M. Pera and W. Shea, eds., *Persuading Science.* Canton, Mass.: Science History Publication, pp. 3–28.

Kuhn, T. [1962] 1970. *The Structure of Scientific Revolutions.* 2d ed. Chicago: University of Chicago Press.

―――. 1977. *The Essential Tension.* Chicago: University of Chicago Press.

―――. 1978. *Black-Body Theory and the Problem of Quantum Discontinuity: 1894–1912.* Oxford: Oxford University Press.

Latour, B. 1987. *Science in Action: How to Follow Scientists and Engineers Through Society.* Milton Keynes: Open University Press.

Levine, J. 1989. "Reaction to Opinion Deviance in Small Groups." In P. Paulus, ed., *Psychology of Group Influence.* Hillsdale: Lawrence Erlbaum Associates.

Manicas, P. 1986. *A History and Philosophy of the Social Sciences.* Oxford: Blackwell.

Polanyi, M. 1957. *Personal Knowledge.* Chicago: University of Chicago Press.

Popper, K. 1957. *The Poverty of Historicism.* New York: Harper & Row.

Proctor, R. 1991. *Value-Free Science? Purity and Power in Modern Knowledge.* Cambridge, Mass.: Harvard University Press.

Rorty, R. 1979. *Philosophy and the Mirror of Nature.* Princeton: Princeton University Press.

Ross, D. 1991. *The Origins of American Social Science.* Cambridge, England: Cambridge University Press.

Science Citation Index. 1961– . Bi-monthly. Philadelphia: Institute for Scientific Information.

Shapin, S., and S. Schaffer. 1985. *Leviathan and the Air-Pump: Hobbes, Boyle, and the Experimental Life.* Princeton: Princeton University Press.

Stehr, N. 1992. *Practical Knowledge.* London: Sage Publications.

Unger, R. 1987. *Politics: A Work of Constructive Social Theory.* Vols. 1–3. Cambridge, England: Cambridge University Press.

White, H. 1972. *Metahistory.* Baltimore: Johns Hopkins University Press.

Wittgenstein, L. 1978. *Remarks on the Foundations of Mathematics.* Cambridge, Mass.: MIT Press.

6

Producing Sunspots on an Iron Pan

Galileo's Scientific Discourse

R. Feldhay
Cohn Institute for History and Philosophy of Science and Ideas, Tel Aviv University

In a letter from 1611 J. Blancanus—a Jesuit mathematician—maintained that the moon was mountainous only at its center. The mountains on the moon did not reach its outer circle, which was "entirely lucid, without any shadow or sign of inequality" (Favaro 1968, vol. 11, 127; my translation). This, according to Blancanus, was demonstrated by observation. Galileo Galilei reacted to Blancanus's contentions, "How, then, do we know that the moon is mountainous? We know it not simply by the senses, but by copying and combining up discourse with observations and sensory appearances" (ibid., 183; my translation). Galileo's usage of the word discourse (*discorso*) is intriguing. Avoiding his frequently repeated formula, according to which knowledge is attained through observations and sense experiences on the one hand, and demonstrations on the other, he emphasized the interdependence between observation and language. In seventeenth-century Italian and French *discorso* or *discours* signified the form of writing gradually replacing the more traditional "commentary" or "question." Since the fourteenth century the Latin word *discursus* had been used to denote "an argument." However, partly due to Galileo's texts, the term *discorso* acquired the meaning of "rational (linguistic) practices" in general, a sense that became obsolete already at the end of the eighteenth century. Facts are not known by simply being observed, Galileo seems to have contended. Observation is always mediated by arguments.

The urge to uncover the dependence of sense evidence on words and arguments (versus the call of sixteenth-century scholastics to return to experience in building up arguments; see Schmitt [1969] 1981, Dear 1987) represents a new insight which echoes through Galileo's text without ever being fully articulated. It may be taken as a symptom signaling a deviation from the traditional mixed mathematical science out of which Galileo's early work had emerged. Galileo's notion of discourse, however, remained implicit and vague, and hence in need of interpretation. I attempt to illuminate it through an analysis of the dispute he had with another Jesuit mathematician, C. Scheiner, on the phenomena of sunspots. Scheiner (1573–1650) studied philosophy and mathematics in Ingolstadt, and taught mathematics and Hebrew there from 1610 to 1616. In 1623 he became Rector of the college of Neisse in Rome. In 1633 he was asked by the emperor Ferdinand II to return to Germany, and then stayed six years in Vienna. In 1639 he returned to Rome, where he remained until his death.

In early 1611, Scheiner observed certain spots on or near the sun, and attempted to account for their location, material essence, and movement. Scheiner had written three letters concerning his observations, which he sent under the pseudonym of "Appelles" to M. Welser. Welser, a widely educated nobleman of Augsburg, who held several political positions, was also a patron of the arts and a member of the Accademia dei Lincei. He sent the letters to Galileo, urging him to reply. Galileo reacted slowly, sending a letter to Welser, which in turn gave rise to a more detailed response from Appelles in the form of three letters named "A More Accurate Disquisition." Galileo's remaining letters to Welser were written in response to Appelles's challenge in both his first letters and in the "Disquisition" (in Favaro 1968, vol. 5, 20–70).

Galileo's insistence on the interdependence of sense data and arguments in the letter to Blancanus offers a clue to his strategies in the dispute on sunspots. First, Galileo's words remind us that those strategies should not be reduced to the mathematical proof he offered for the contiguity of the spots with the sun. His own emphasis on discourse urges a search for the specific linguistic practices by which he sought to incorporate the newly discovered phenomena of sunspots within a system of knowledge that could be differentiated and legitimized in accordance with his needs and interests.

I suggest three categories of practices through which Galileo's scientific discourse can be analyzed. The first category is concerned with the constitution of scientific objects, and includes three types of strategies by which observations were objectified. A thing became an object of science by its enunciation in mathematical language, by specifying its conditions of visibility, and by imagining it as being given to manipulation. Thus, Galileo's proof of the contiguity of the spots with the sun represents but one aspect of a much wider network of practices which he applied to the phenomena in order to incorporate them into his discourse.

The second category of practices relates to the construction of boundaries by which Galileo attempted to differentiate his discourse from other systems of knowledge in the same cultural field (see Bourdieu 1984; 1993), especially the mixed mathematics practiced by his Jesuit interlocutors.

The last category of practices by which Galileo's discourse was differentiated involves techniques of establishing the speaker's authority, in this case the authority of a philosophical astronomer. Had Galileo been content with presenting his proof of the contiguity of the spots with the sun, the authority of a traditional astronomer would have sufficed. Galileo's mathematical proof represented no innovation in the traditional practices of astronomers. Because of the broader conception of the astronomer's role as a codebreaker of nature's language, Galileo needed to revert to complicated rhetorical techniques by which he both interpreted the significance of sunspots, and simultaneously concealed the role of interpretation in his science.

My arguments are hermeneutical. I have presupposed a notion of discourse (see Foucault 1972)—a network of practices by which scientific objects are constituted, boundaries are constructed, and the authority of the speaker is established. My contention is that science is basically a discourse. This does not mean that science is only a discourse. Rather it means that science has its own strategies for treating natural objects, within a well-defined discursive space, and by certain authorized speakers. The essence of science cannot be defined a priori as that which is based on proof, not on persuasion. Whatever makes it sui generis must be established, among other things, by a close reading of the texts in which scientific theories exist, not by a presupposed distinction between proof and rhetoric.

The Constitution of Sunspots as Objects of Scientific Discourse

The nature of the spots as objects of science did not follow automatically from the observation of several astronomers in 1611. Galileo's letters on sunspots contain many traces of their first interpretation by Scheiner. But only a reading of Scheiner's texts reveals how broad was the factual background common to his interpretation and to that later suggested by Galileo. According to both, sunspots were no illusions (Scheiner 1968, 24; Drake 1957, 91). They were real, material objects indicating true motions in heaven. The practices of establishing the existence of such phenomena were likewise similar for both Galileo and the Jesuit. Being aware of the exclusiveness of telescopic phenomena (see Dear 1987, 156) they both insisted on the continuous and repetitive nature of their observations and recruited witnesses to confirm their statements (Scheiner 1968, 25; Drake 1957, 91). Both devoted much space to detailed descriptions of their mode of observation, the material means available for the purpose, and the correct way of using the necessary instruments. Both were concerned with specifying the particular techniques of representing the results of observations. Moreover, the two astronomers also agreed that the spots traveled across the sun's disk in approximately fifteen days; that they became smaller around the edge of the sun; that they were seen to unite and divide, constantly changing their shape as they moved around the sun; and that they moved faster in the middle of the sun than on the perimeter (Scheiner 1968, 29–30; Drake 1957, 91).

Recognition of the factual background common to Galileo and Scheiner also draws attention to the Jesuit's categories for discussing sunspots, which varied little from Galileo's: Location, motion, shape, and size were primary, and required mathematical language as a basic mode of enunciation. This comes as no surprise, for Galileo and Scheiner were both deeply immersed in the long traditions of astronomy as a mixed mathematical science, dealing with (physical) celestial phenomena by mathematical methods of investigation. Yet, Galileo's sunspots differed significantly from Scheiner's. Scheiner contended that the spots were shadows of real stars projected upon the sun's surface while eclipsing it, and moving, like all celestial bodies, "by their own movements" (Scheiner 1968, 26; my translation), close to the sun but not contiguous with it. Scheiner further insisted that the spots were not clouds (ibid., 30). Their blackness indicated a much more solid, opaque, and dense substance than clouds, namely, that of stars. Toward the end of his third letter Galileo summed up his verdict on sunspots, their substance, location, and motion, "I have demonstrated that the sunspots are neither stars nor permanent materials, and that they are not located at a distance from the sun but are produced and dissolved upon it in a manner not unlike that of clouds and vapors on the earth" (Drake 1957, 143). A detailed examination of Galileo's proof of the spots' contiguity with the sun in its context signals two features by which Galileo's discourse is differentiated from that of Scheiner's, who remained within the tradition of the mixed sciences even though he contributed to its modification.

Scheiner gave a general description of the spots' size, speed, and change of shape, deeming them unfavorable to the hypothesis that they were on the sun, but chose not to pursue the matter further. Galileo proceeded to actually measure the properties of the spots, and was able to deduce from his measurements the accurate location of the spots on the sun. This success in terms of the traditional methods of the mixed sciences, however, necessitated Galileo's more radical claims, namely, that the spots indicated the corruptibility of heavenly matter, in analogy with earthly matter. Thus, when radically pursued, the enunciation of the spots in mathematical terms led Galileo to examine the visible motion of the spots in analogy with the invisible motion of the earth, and even to speculate on the spots' material conditions through their imaginary manipulation. What finally constituted the spots as scientific objects was not simply the geometrical proof of their contiguity with the sun, but also the ability to guide the observer's gaze and sense of touch in order to see and feel what was not observable and touchable.

Fixing the sun as the frame of reference within which the spots should be discussed was Galileo's first target. First he located them on a network of coordinates, which he drew on the surface of a sphere, thus making possible the measurement of their breadth and length. Observation showed, he argued, that whereas the spots tended to have the same length wherever they appeared on the sun, their breadth changed as they moved from the edges to the central parts of the sun's disk, being very little at the moment of appearance and disappearance and growing toward the center. Such behavior could best be explained by the geometrical rules of perspective, which described the fixed length and expanding breadth of an object drawn on a spherical surface as "foreshortening." The point of maximum foreshortening, he explained, was also the point of maximum thinning, a fact which indicated that the spots must be located on the sun, or at an imperceptible distance from it; otherwise, maximum foreshortening would occur outside the face of the sun (ibid., 107–08). Two additional mathematical arguments showed that the spots must be on the sun, not at a distance from it: One concerned the decreasing distances they traveled in the same time as they approached the edge of the sun, and the other exposed the mode of variation in the distances between one spot and another, which could be accounted for only by circular motion on the sun (ibid., 108–09).

The successful location of the spots immediately raised two problems. The first concerned the source of their motion. Galileo referred this motion to the rotating motion of a spherical sun, or its ambient. Galileo, however, lacked any physical principle from which such celestial motion could be deduced, apart from Copernicus's speculation which connected rotational movements to a spherical shape. Not surprisingly, perhaps, his argument for the rotational motion of the sun or its ambient—indicated by the spots—came from his experience with the motion of physical bodies upon the earth. The sun's motion was described in terms of "inertia," of which he had learned by watching the motion of physical bodies on earth, and the principle of which he here formulated for the first time:

For I seem to have observed that physical bodies have physical inclination to some motion (as heavy bodies downward), which motion is exercised by them through an intrinsic property and without need of a particular external mover, whenever they are not impeded by some obstacle. And to some other motion they have a repugnance (as the same heavy bodies to motion upward), and therefore they never move in that manner unless thrown violently by an external mover. Finally, to some movements they are indifferent, as are these same heavy bodies to horizontal motion, to which they have neither inclination (since it is not toward the center of the earth) nor repugnance (since it does not carry them away from that center). And therefore, all external impediments removed, a heavy body on a spherical surface concentric with the earth will be indifferent to rest and to movements toward any part of the horizon. And it will maintain itself in that state in which it has once been placed; that is, if placed in a state of rest, it will conserve that; and if placed in movement toward the west (for example), it will maintain itself in that movement. (Ibid., 113)

The sun, having neither a disinclination nor a propensity for rotational motion, could participate in such a motion of the ambient, or continue forever in such a motion of its own, just as a ship would continue to move around the earth forever once it were set in motion and not arrested by external impediments (ibid.).

Galileo's explanation of the sunspots' motions seems to be susceptible to two contradictory interpretations.

1. Galileo's achievement may be perceived as an early formulation of the first modern law of nature, the law of inertia, which became the conceptual matrix of both his Copernicanism and his science of mechanics, a law from which he could deduce a common explanation for celestial and terrestrial motions. Such an interpretation, however, is only possible by hindsight, at the risk of distorting any genuinely historical account of Galileo's discourse, for the concept of circular inertia never became a general principle for Galileo, nor did ne ever attempt to deduce observed motions from it. In fact, many years later, in the *Dialogue* of 1632 Galileo admitted that he did not have a general explanation of motion:

But we do not really understand what principle or what force it is that moves stones downward, any more than we understand what moves them upward after they leave the thrower's hand, or what moves the moon around. We have merely, as I said, assigned to the first the more specific and definite name "gravity", whereas to the second we assign the more general term "impressed force" (virtu impressa), and to the last-named we give "spirits" (intelligenza) . . . and as the cause of infinite other motions we give "Nature". (Galilei 1962, 234–35)

This does not mean that Galileo was skeptical of the ability of the human mind to achieve causal explanations. It means, however, that the concept of circular inertia did not acquire a theoretical status in Galileo's discourse (Koyré 1978, part 3, chap. 3), despite that it had a function both in the physical account of free fall as well as in the refutation of Ptolematic arguments against the possibility of the earth's motion.

2. Alternatively, Galileo's concept of circular inertia may be interpreted in terms of the "natural motions" which play such an important role in the Aristotelian cosmos. In fact, Galileo explicitly calls the unaccelerated circular motion of bodies moving on a horizontal plane "natural motion," and contends that it is the only natural motion that

exists or can exist. Nevertheless, Galileo's concept of natural motion differs from Aristotle's, mainly because it explains nothing. Natural motions, according to Galileo, are nothing but names of phenomena for which we have no explanation. In the Aristotelian universe, natural motion is an expression of the essence of natural bodies: A heavy body "naturally" falls downward, a motion caused by its essential "heaviness." Galileo vehemently rejected this type of explanations, and dedicated the whole first day of the *Dialogue* to their refutation. In his third letter on sunspots, he had already articulated his reason for this rejection:

For in our speculating we either seek to penetrate the true and internal essence of natural substances, or content ourselves with a knowledge of some of their properties. The former I hold to be as impossible an undertaking with regard to the closest elemental substances as with more remote celestial things . . . . I know no more about the true essences of earth or fire than about those of the moon or sun, for that knowledge is withheld from us, and is not to be understood until we reach the state of blessedness. (Drake 1957, 123–24)

Moreover, the gap between Aristotle's concept of natural motion and Galileo's is most acutely expressed in the fact that Aristotelian natural motions are *visible* most of the time. Galileo's unaccelerated circular motion of bodies moving in a plane concentric to the earth is, in contrast, *unobservable*.

Galileo's concept of circular inertia, which appears for the first time in the "Letters on Sunspots" (ibid.) cannot be reduced to a modern law of nature; neither is it reducible to an Aristotelian natural motion. Galileo's discursive strategy is significant as a clue for exposing a recurrent feature of his discourse, "For I seem to have observed that physical bodies have physical inclination to some motion . . ." (ibid., 113). He does not formulate a principle, but rather chooses to tell a story. Observation of the motion of heavy bodies downward and upward somehow led him to something which is wholly unobservable, but which he describes as continuous with the observable, namely, to the ever continuous motion on a horizontal plane—which is truly spherical—concretized by the motion of a ship forever circling the earth. What looks as the horizon is in fact spherical, but unobservable as such. However, only the passage from the observable to the unobservable enables an argument based on an analogy: between the horizontal-circular motion on earth, and the rotational motion of the sun.

Galileo's explanation of the spots' motion exposes the large gap that opened between his ability to deduce the location of the spots on the sun mathematically, and his inability to use such mathematical deduction as a physical account of the motion. This gap was erased by the analogy between the "indifferent" motion of heavy bodies on earth and the rotational motion of the sun inferred from the phenomena of sunspots.

Scheiner, who considered the visible motion of the spots in terms of the motion of invisible stars, argued that those stars had their proper motion like all other celestial bodies. Galileo, who argued from the inner position of a Copernican, ran the risk of exasperating the weakness of his theory in accounting for the invisible motion of the earth by having to assume another invisible motion, that of the rotating sun. The analogy between the inertial motion of heavy bodies on earth and the inertial motion of the sun or its ambiance thus served to save both difficulties by using the visible (sunspots' motion) to justify the invisible motions implied by his theory.

The second problem raised by the proof of the spots' contiguity with the sun concerned the spots' failure to return after fifteen days, although they were clearly seen to cross the sun's disk in approximately that period. Scheiner considered this fact as the ultimate proof that the spots were *not* on the sun (1968, 26). Galileo deemed it an argument for their generation and dissolion on the sun, and used an even more audacious analogy to substantiate his claim. He contended that the spots resembled clouds:

[If], proceeding on the basis of analogy with materials known and familiar to us, one may suggest something that they may be from their appearance, my view would be exactly opposite to that of Apelles . . . . I find in them nothing at all which does not resemble our own clouds. (Drake 1957, 98)

He described exactly how the analogy worked in terms of the periods of generation and decay, condensation, expansion, shape, color, size, opacity, number, and the like.

At first, the analogy seemed innocent. It was tentatively suggested, and was accompanied by repetitious remarks about our inability to know, strangely untypical of Galileo's polemical style (e.g., "I see nothing discreditable to any philosopher in confessing that he does not know, and cannot know, what the material of the solar spots may be" [ibid.]). Its radical significance was soon revealed, however, when Galileo further suggested that the sun and its spots resemble the earth and its clouds if the earth is imagined as having light of its own:

[T]here is no doubt that if the earth shone with its own light and not by that of the sun, then to anyone who looked at it from afar it would exhibit congruent appearances. For as now this country and now that was covered by clouds, it would appear to be strewn with dark spots that would impede the terrestrial splendor more or less according to the greater or less density of their parts. These spots would be seen darker here and less dark there, now more numerous and again less so, now spread out and now restricted; and if the earth revolved upon an axis, they would follow its motion . . . . In a word, no phenomena would be perceived that are not likewise seen in sunspots. (Ibid., 99)

Apparently, a rotating earth and its clouds were presented as a model for understanding the sun and its spots. Sunspots generate and dissolve like clouds, and are carried by the motion of a body rotating on its axis. In fact, however, since the rotation of the earth on its axis could not be deduced from any physical principle, the observation of the movement of sunspots and their interpretation by means of such a model was a way of arguing for the unobservable motion of the earth.

Galileo's explanation of the spots' motion, as well as his account of their substance, led him to invoke "invisibles," which were somehow justified by observed facts. Such invisible entities were not unknown to traditional astronomical discourse. Astronomers have always attempted to account for the observable (e.g., retrograde motions of the planets, or their unequal velocities) in terms of the unobservable motion of epicycles and eccentrics. These unobserved entities, however, remained qualitatively different from observed ones. Their status as legitimate objects of science did not depend upon their visibility. Almost no one ever attempted to put them on a par with observed objects. Galileo's strategies differed radically from traditional astronomers as he systematically strove to create a continuity between the observable and the unobservable through analogical arguments invoked to erase the gap left by his mathematical demonstrations.

In Galilean discourse a thing was constituted as a scientific object by mathematical enunciation and by being given to observation. But a further condition was required: Scientific objects should in some sense be amenable to manipulation, even if only by analogy. The second letter to Welser includes a technique by which sunspots could be regarded as "manipulable":

I liken the sunspots to clouds or smokes. Surely if anyone wished to imitate them by means of earthly materials, no better model could be found than to put some drops of incombustible bitumen on a red-hot iron plate. From the black spot thus impressed on the iron, there will arise a black smoke that will disperse in strange and changing shapes. And if anyone were to insist that continual food and nourishment would have to be supplied for the refueling of the immense light that our great lamp, the sun, continually diffuses through the universe, then we have countless experiences harmoniously agreeing in showing us the conversion of burning materials first into something black or dark in color. Thus we see wood, straw, paper, candlewicks, and every burning thing to have its flame planted in and rising from neighboring parts of the material that have first become black. It might even be that if we more accurately observed the bright spots on the sun that I have mentioned, we should find them occurring in the very places where large dark spots had been a short time before. (Ibid., 140)

The strange, changing spots produced on an iron pan were *like* the strange, changing shapes of sunspots, contended Galileo. The conversion of burning materials from something dark to glorious illumination was likened to the imagined burning of materials on the surface of the sun. It was even deemed possible that the bright spots appeared exactly in the place of preexisting dark spots, just like the flame which appeared after the burning of bitumen dropped on an iron pan.

Mathematical argumentation enabled Galileo to prove the contiguity of sunspots with the sun. Their appearance and disappearance could thus be interpreted as demonstrating the corruptibility of the heavenly region. Such indirect proof, however, was too abstract for Galileo. Only through an analogy between the manipulation of earthly matter and the mutability on the sun was the argument of corruption validated and the qualitative difference between the heavens and the earth delegitimized:

[In] that part of the sky which deserves to be considered the most pure and serene of all—I mean in the very face of the sun—these innumerable multitudes of dense, obscure, and foggy materials are discovered to be produced and dissolved continually in brief periods. Here is a parade of productions and destructions that does not end in a moment, but will endure through all future ages, allowing the human mind time to observe at pleasure and to learn those doctrines which will finally prove the true location of the spots. (Ibid., 119)

The dispute on sunspots reveals structural differences between Galileo's discourse and Scheiner's, although both went beyond traditional astronomy, relying on sense evidence and mathematical proofs. Like

Galileo, Scheiner was committed to the ontological reality of the spots and conceived of astronomy as a discourse of truth. Moreover, even traditional cosmology did not enjoy an absolute canonic status for Scheiner, but was subject to modification. Scheiner actually thought that from the motions of sunspots he could learn about the motion of Venus and Mercury around the sun (1968, 26). However, the motion of the earth seems to have been unpalatable for the Jesuits, either because of the difficulty it presented to biblical exegesis or because it could not be accounted for in terms of the existing physics. Being problematic both theologically and philosophically, the Jesuits, and Scheiner among them, chose to ignore it. Consequently, Scheiner resorted to the traditional astronomers' strategies of constituting the objects of astronomy, who used to invoke invisibles for explaining observed facts, without committing themselves to the nature of those explanatory invisibles. According to Scheiner the spots were shadows of invisible stars projected upon the sun's surface at conjunction. Scheiner, however, stopped short of committing himself to that interpretation and left it open for the criticism of philosophers.

Galileo, following Copernicus, presupposed the motion of the earth, although he still lacked a full-fledged physical theory that would justify his conviction. To suppose the motion of the earth, however, meant placing an unobservable in the center of astronomical discourse. Epicycles and eccentrics—the traditional unobservables in astronomy—had always played an explanatory role, never that of the thing to be explained. To place an unobservable as an explanandum meant it could be legitimated only by observables. The structure of traditional astronomy was violated by such an act. Instead of focusing on techniques for making observables (the unequal motion of the planets) understood in terms of unobservables (eccentrics and epicycles), it was now necessary to make the unobservable (the motion of the earth) comprehensible in terms of the observable (motion of sunspots). Sunspots seemed to provide Galileo with an outstanding opportunity for justifying the unobservable motion of the earth in terms of the observable motion of the sun if the analogy between the two could be made manifest to the senses. Galileo's main effort in the letters was not dedicated to the geometrical demonstration of contiguity—which was perfectly in tune with the traditions of the mixed sciences—but rather to elaborating the analogies between the sun's rotational motion and the earth's, between sunspots and clouds, and between earthly and heavenly matter in order to make the unobservable in some sense observable, and the immutable in some sense manipulable.

The Construction of New Boundaries in Galileo's Discourse

No less significant was the challenge sunspots presented to the boundaries of astronomy, and the reactions to this challenge. First, a few words on the broader institutional context of the dispute on sunspots. The prolific educational activity of the Jesuits gave rise to a more-or-less differentiated cultural field, whose center was the Jesuit intellectual establishment and in which university astronomers and philosophers (Magini in Bologna, Cremonini in Padua, Lagalla in Rome), court mathematicians and philosophers (Kepler in Prague, Galileo in Florence), and Jesuit astronomers and philosophers took a position as speakers (Feldhay 1995). The *Ratio studiorum* of 1599 (see Salmone 1979)—the official document in which the Jesuit idea of the organization of knowledge was formalized—represents the precise points on the intellectual map where the Jesuit authorities deliberately took action to preserve boundaries.

The *Ratio* endorsed the boundaries between mathematics, natural philosophy, and theology. Such endorsement expressed an official commitment to the Thomistic organization of knowledge (see Feldhay 1987). Thus, the Jesuits' break with the contemplative inclinations of Thomism toward a more liberal Thomistic theology and practical science was counterbalanced by their avoidance of openly challenging the boundaries dictated by Thomism and guarded by the Church's old, established intellectual elite. C. Clavius's program for a radical reform of the educational system was rejected, and an institutional constraint on astronomical discourse, confining it to the mixed mathematical sciences, was firmly established. This did not mean that the discussion of physical problems by astronomers was eradicated. It did mean, however, that the system was sensitive to transgressions of the boundaries between mathematics and physics, and monitored them closely.

The boundaries of scientific discourse always reflect two sets of constraints: One set is imposed by the objects of discourse, the other by the position of other speakers in the cultural field and their interests. Thus the boundaries mediate between the inner core of discourse and the environment in which it is embedded. Their construction represents the limit of possible arguments inside a discourse and of legitimate exercise of power from the outside. The reconstruction of Scheiner's discussion of "substances" within an astronomical text illuminates the gap between the discursive rules which operated in the Jesuit astronomical discourse and those operating in Galileo's, with his explicit challenge to the boundaries between mathematics and philosophy.

Scheiner's position may be understood in terms of transgression as it has been described by M. Foucault:

> It is like a flash of lightning in the night which, from the beginning of time, gives a dense and black intensity to the night it denies, which lights the night from the inside, from top to bottom, and yet owes to the dark the stark clarity of its manifestation, its harrowing and poised singularity; the flash loses itself in this space it marks with its sovereignty and becomes silent now that it has given a name to obscurity. (1977, 35)

The ambiguous nature of transgression probably requires a metaphorical language if it is to be expressed with any accuracy: transgression as a momentary illumination of an order of things, immediately swallowed by the authoritative hegemony of the very order which it defies.

Scheiner's letters on sunspots and his *Disquisition* were an attempt to determine the motion, location, and substance of sunspots. The investigation of the movement and location of celestial bodies was a standard preoccupation of astronomers, performed through geometrical representation of observations, calculations, deductions, and analogical arguments. Whereas the greater part of Scheiner's letters was devoted to such legitimate deliberations, still the third part of the third letter was wholly concerned with the substance of sunspots. Such a discussion was clearly beyond the boundaries of legitimate astronomical discourse. "What are they (these spots) finally?" Scheiner asked, and proceeded to give an answer, regardless of boundaries.

First, he chose to deal with what sunspots were not: They were not clouds. The legitimacy of this statement was guaranteed a priori, for clouds were an earthly substance, and by denying the "cloudiness" of sunspots Scheiner denied their earthly nature. He then moved to speak in the name of the philosophers, who would surely agree, he said, that sunspots must be "parts of some heaven" (Scheiner 1968, 30; my translation), or, in other words, stars. Only finally did he arrive at what he thought must be the conclusion of the astronomers, "That they are bodies existing in themselves, solid and dark, and for this reason alone they must be stars, no less than the moon and Venus, which appear black in the part turned away from the sun" (ibid.; my translation).

He was aware that philosophers would deduce the heavenly nature of the spots from the nature of the heaven, of which they were part. (He was alluding, here, to the traditional cosmography whereby observed celestial bodies like the planets were carried by material spheres or heavens, filling every spot in space.) In contrast, the astronomers' point of departure was phenomena—the observation of dark, shadow-throwing bodies. Such observed "facts" allowed the astronomers to conclude—independently of philosophical speculation on the nature of the heaven—that they were substances "bodies existing in themselves," and to draw an analogy between spots and other stars. Thus, the astronomers' approach was kept different and separate form the philosophers'. Astronomers, Scheiner seems to say, observe and interpret their observations. Phenomena *indicate* (in the literal sense of the word, viz., *point to*) something for the astronomer, although astronomers are not supposed to be preoccupied with the essence of the phenomenon observed and with its logical relation to other essences.

Scheiner's act of transgression consisted in his audacity in speaking openly about substances beyond the domain of astronomers, and within that of philosophers. Scheiner, however, was far from defying the boundaries between the two disciplines imposed upon the system. He took care to neutralize the effect of his transgression by insisting that astronomers talk differently from philosophers about substances, thus leaving the latters' discursive space free from intrusion, and their authority untouched. The question he asked about sunspots, "What are they?", was subverted by him into a question of another kind, namely, "What do they signify?" Instead of proceeding from effects (spots) to causes (stars) as a philosopher would do, he interpreted the effect as a sign pointing to, or indicating, a phenomenon, which was then woven into a network of other phenomena (the motion of Venus and Mercury, conjunction with the sun, the order of the planets' orbits, and such) all within the legitimate domain of astronomers. Finally, Scheiner never let his reader forget that real knowledge could only stem from an agreement between philosophers and astronomers, confirming, in fact, the superior position of the philosophers in the disciplinary hierarchy. Until the philosophers' quest for proof was satisfied, no knowledge could be truly legitimized, no "fact" could be recognized as certain:

It occurs to me that between the sun and Mercury and Venus, in due distance and proportion, there revolve very many wandering stars, of which only those which run into the sun in its motion become known to us. If it so happens—and I have not yet completely abandoned the idea—that we could contemplate only those very stars which are close to the sun, this dispute could then be settled. (Ibid.; my translation)

The nature of sunspots, then, was not a matter that could be determined yet, either by an individual author, or even by the community of astronomers. It was still left to discussions between astronomers and philosophers.

Scheiner's texts reflected institutional constraints even at the moment of transgression. The distinction between mathematics and philosophy in terms of their different methodologies and different themes was maintained even while the methods and objects within each domain were undergoing change. Astronomy was a mixed science subalternated to mathematics, not wholly autonomous, and unable to objectify its phenomena independently. This could only be done in cooperation with philosophers, charged with the responsibility of determining substances and their natures, through complex negotiations by which truth would finally shine forth in the form of consensus.

The boundaries reproduced in Scheiner's texts provoked Galileo's most severe criticisms of the Jesuit, even though the former's transgression did not pass entirely unnoticed. In fact, Galileo used Scheiner's assertion of the unpassable limit—the consensus among philosophers and mathematicians—in order to declare the necessity of its destruction. Galileo first attacked the accepted boundary between astronomy and natural philosophy by remarking upon Scheiner's philosophical interests, "Yet I seem to see in Appelles a free and not a servile mind. He is quite capable of understanding true doctrines; for, led by the force of so many novelties, he has begun to lend his ear and his assent to *good* and *true* philosophy" (Drake 1957, 96; emphasis added). Scheiner was thus praised for his transgression and encouraged to enter into the sphere of the philosophers. Galileo then strove to transform this transgression into a revolutionary act. In this spirit he made his famous distinction between *mathematical astronomers*—who utilized fictive inventions like eccentrics, deferents, equants, and epicycles in order to facilitate their calculations—and *philosophical astronomers*—who went beyond the attempt at "saving the appearances" and investigated the true nature of the universe. For the first time Galileo defined the role of the new astronomer in terms of a vocation embodying a set of noble values, "For such a constitution exists; it is unique, true, real, and could not possibly be otherwise; and the *greatness* and *nobility* of this problem entitle it to be placed foremost among all questions capable of theoretical solution" (ibid., 97; emphasis added).

After reading Scheiner's second series of texts Galileo realized to what extent the Jesuit was confined within the traditional limits of his profession. In response, Galileo ridiculed philosophers ("it is vain to run to such men asking for support" [ibid., 135]) and delegitimized *traditional* mathematicians dealing with astronomical problems, "As to the mathematicians, I do not know that any of them have ever discussed the hardness and immutability of the sun, or even that mathematical science is adequate for proving such properties" (ibid.). He no less vehemently rejected the consensus among philosophers and mathematicians, "Any adversary would resolutely reject . . . the proof Apelles adduces for it, which is that such is the prevailing opinion (according to him) among philosophers and mathematicians" (ibid., 134).

Galileo's and Scheiner's texts reflect the contradictory interests involved in the boundaries between astronomy as a mixed science subalternated to mathematics and natural philosophy as a higher discipline aspiring to demonstrated truth. The boundaries implied in Scheiner's texts were imposed by institutional requirements according to which astronomy was conceived as a subalternated science whose capacity to provide causal knowledge about nature was doubtful. Galileo rejected the boundary between mathematics and natural philosophy, and created the ideal of philosophical astronomy. In doing so he used his authority as the court philosopher and mathematician to Cosimo II. However, this position also pushed him to emphasize the philosophical implications of his work in a manner which caused difficulty to the Jesuits. While Scheiner was obliged to keep the distinction between mathematics and physics by recognizing the preeminence of the philosophers, Galileo was bound to blur it, both explicitly and implicitly. Explicitly, he did so by dividing astronomers into two groups, mathematical and philosophical, thus facing Scheiner with an impossible choice; and he did it implicitly by insisting on a discussion of sunspots in analogy with clouds, namely, with earthly materials.

Galileo's distinction between mathematical and philosophical astronomers left the Jesuit mathematicians no space within that classification, for mathematical astronomers, according to that distinction, only "supposed" the objects of their discourse for purposes of calculations. They had no claims upon reality. That, of course, did not apply to Scheiner who clearly aimed at truth, and was not satisfied with a hypothetical status for his conclusions. Philosophical astronomers, on the other hand, were bound to investigate the structure of the universe. No challenge was as clear, not only to the boundaries between philosophy and mathematics but also to the hierarchy between them, than this declaration of Galileo, in which he literally dethroned the philosophers from their position, claiming to be their legitimate heir. No Jesuit mathematician could have dreamt of identifying with such an assertion. Thus, Galileo excluded the Jesuits from the cultural field which they both shared long before they excluded him from the place he aspired to occupy with their cooperation.

Authority and Authorization in Galileo's Scientific Discourse

The claim that Galileo's ideal of philosophical astronomy was created in the context of his position at court does not mean that he ignored the need to anchor knowledge in the scientific norms of his age. The external pressure to present his ideas as the product of philosophical astronomy, and thus to differentiate his professional identity from that of practical mathematicians (see Biagioli 1993, chap. 1), only enhanced the need to justify the certain and true status of his contentions epistemically. The "Letters on Sunspots" reveal three stages in the attempt to establish and justify the new "facts."

First came the geometrical proof of the contiguity of the spots with the sun. The proof obviously justified the claim that his hypothesis perfectly saved the phenomena. Galileo, however, did not consider such a procedure satisfactory as a true demonstration, obvious in the light of the following passage that appears in the second letter, immediately after the presentation of the proof:

And just as all the phenomena in these observations agree exactly with the spots' being contiguous to the surface of the sun, and with this surface being spherical rather than any other shape, and with their being carried around by the rotation of the sun itself, so the same phenomena are opposed to every other theory that may be proposed to explain them. (Drake 1957, 109)

Lacking any general principle from which the motion of the spots on the sun and the corruptibility of the heaven could be deduced, the only strategy for claiming the status of scientifically proven knowledge for his conclusions was to eliminate all other possible hypotheses. This was exactly the strategy Galileo used. In most things he followed in the footsteps of Scheiner, eliminating some of the possibilities rejected by the Jesuit as well (e.g., could they be in the air? could they be in the orbit of the moon? and such) although not always for the same reasons. After a few eliminations of this kind, including the rejection of Scheiner's hypothesis, Galileo concluded, "[It] would be a waste of time to attack every other conceivable theory" (ibid., 111).

From the logical point of view, Galileo's effort to gain the status of scientific truth for his hypothesis was doomed to failure, however. Logically speaking, new hypotheses could always be derived that would "save the appearances" better than his own. Hence, his assertion that it would be a waste of time to attack all other theories was an ambiguous rhetorical gesture. While all possible hypotheses had supposedly been eliminated and the true one allowed to emerge and be scientifically established, it in fact revealed the inaptitude of this method in the study of nature. It had now become clear that there could be no legitimate transition from the status of a probable hypothesis to the status of a scientifically established truth through an accumulation of empirical evidence. At this stage the expectation that "mathematical astronomy" would become "philosophical astronomy" through more "evidence of things" had to be modified. The rest of the second letter represents a shift in Galileo's strategies. It contains the first enunciation of the principle of "indifferent" motion, testifying to Galileo's recognition that only by discovering a general principle of motion from which the rotation of the sun would be deduced could the appearance be established as scientific facts.

Galileo's principle of "indifferent" (i.e., inertial) motion, as he admitted, was based upon observation and analysis of the motion of heavy bodies on the earth. Also, the example of a ship "having once received some impetus through the tranquil sea, would move continually around our globe without ever stopping" (ibid., 114) was invoked in order to legitimate the sun's rotation in terms of an analogy to an earthly body moving on a spherical surface. In the absence of a clear concept of force and gravitation, however, Galileo could not yet establish "indifferent"—inertial—motion as the fundamental principle of mechanics. Thus, the validity of his argument about the rotation of the sun wholly depended upon the analogy he claimed between the "indifferent" motion of earthly physical bodies and the motion of celestial bodies. This analogy was based on an act of interpretation: his reading of spots in terms of clouds, and treatment of the relationship between the spots and the sun as being similar to that of the relationship between the clouds and the earth. In fact, he was caught in a hermeneutical circle, for the analogy between the moving spots and the sun on the one hand, and a moving earth with its clouds on the other, had already presupposed the motion of the earth without any argument. Paradoxically, Galileo's attempt to step out of the framework of the mixed mathematical sciences toward the formulation of a general principle of motion from which the motion of sunspots could be deduced led him to base his reasoning on an analogy inspired by rhetoric rather than based on scientific proof.

The shift from "proof" to "analogy" had far-reaching implications for the authority of the scientist in Galileo's discourse. A proper "demonstration" was in no need of any special authority in order to be accepted as "truth." A "correct analogy," on the other hand, could have no claim to truth apart from the speaker's authority. Thus, Galileo was forced to resort to some radical strategies in order to justify his position as the discoverer of the "true analogy."

From his first letter Galileo represented himself as an individual standing alone against a crowd of persecutors, "Even the most trivial error is charged to me as a capital fault by the enemies of innovation, making it seem better to remain with the herd in error than to stand alone in reasoning correctly" (ibid., 90). At this early stage of his Copernican campaign, this self-representation was remote from reality. Galileo had just been cordially received by the mathematicians of the Roman College, who confirmed most of his discoveries. Welser had crowned him as a leader, "You have led in scaling the walls, and have brought back the awarded crown . . . . I hope you will be pleased to see that on this side of the mountains also men are not lacking who travel in your footsteps" (ibid., 89). The Linceans were at his side, expressing the absolute loyalty demanded from them by Prince Cesi, "Every one of us will always write on your behalf" (Favaro 1968, vol. 11, 282; my translation) he reassured Galileo in a letter of March 1612. A large number of letters from other sympathizers, Cardinal Barberini (see ibid., 317–18, 325), Cardinal Francesco di Joyeuse (see p. 373), and the Aristotelian philosopher Lagalla (see pp. 357–59) most prominent among them, show that he was not as isolated as he pretended to be.

Even more problematic was Galileo's claim to be right on account of his "correct reasoning." As argued, Galileo was particularly eager to prove his theories, and was not content with the status of probability they could secure in the framework of the mixed sciences. Only a proof could justify his discourse as philosophical, and not just mathematical astronomy. His deduction of the motion of sunspots from a general principle of motion—the only way to legitimate his hypothesis scientifically—depended, however, on the implicit analogy between the motion of earthly bodies and celestial bodies, whose laws of motion were as yet completely obscure. The analogy he suggested was a choice arrived at in the context of the new paradigm suggested by Copernicus, but without much knowledge of a physics of the heavens. It did not stem only from correct reasoning.

Galileo's words, however, should not be taken at face value. Rather than reflecting a cultural reality, they expressed a discursive need to legitimize the individual vis-à-vis the consensus. The dichotomy between the righteous individual and the ignorant masses never existed. The opposers were neither ignorant nor wholly against innovation. Rather, it was a fabricated dichotomy, providing the basis for the Galilean myth which was invented in the text of its protagonist. The myth had a function, however. It was meant as an answer to Scheiner, who did not deny the interpretive character of his statements and tended to base the authority of the astronomer on the consensus of the community of astronomers and philosophers. Against that strategy Galileo posited the authority of the individual philosophical astronomer over and above the consensus of the community, and emphasized the superior moral value of his position.

The most succinct formulation of the authority of the scientist in Galileo's discourse appeared in the context of his reply to the *Disquisition* (i.e., in the third letter to Welser). Scheiner summarized three arguments in support of Venus's motion around the sun: the consensus among philosophers and mathematicians, the analogy between the expected spots of Venus and other spots (i.e., those seen by Kepler who described them as the shades of Mercury), and the "evidence of things," which he called "experience," and of which the most outstanding example was Galileo's account of the phases of Venus (see ibid., vol. 5, 46). Scheiner explicitly stated what had already been clear from his previous three letters, namely, that the source of authority in his discourse was the consensus of the community of observers, practitioners, mathematical astronomers, and philosophers.

Scheiner's sociological insight into the dynamics of scientific communities did not mean that in Jesuit culture there were no canons of proof, including coherent criteria of internal evidence. Neither did it mean that canonic texts were the only authoritative source of knowledge. Rather, it indicated the high degree of institutionalization that characterized Jesuit science. In that context, scientific activity became collective activity. The collective dimension of his enterprise found its most emphatic expression in the language used in the letters. There, all reports of observations were related in the first person plural, "They (the appearances) provided not only me myself, but also my friends, at first with great astonishment, and thereafter with great delight of the soul" (1968, 25; my translation) and "I myself and one of my friends directed together the optical tube toward the sun. . . . We returned however to this matter last October, and again we perceived those spots. . . . We have consulted the eyes of the most various people (ibid.; my translation). Also, the conclusions were presented as the result of collective reasoning:

We have concluded therefore that the imperfection was not in the eyes . . . it was unanimously and correctly concluded by very many people, that these appearances could by no means be situated in the sun because of a defect of the eye . . . and we concluded together and correctly that in this mater the tube was justly to be cleared of any blame. (Ibid., 25–26; my translation)

Compared to Scheiner's collective persona, Galileo's individual "I" was indeed remarkable. True, Galileo too invoked witnesses to confirm his observations. But even the observations he had performed in front of an audience were arranged as a scene in which Galileo showed the others what they should see, "I have observed them for about eighteen months having shown them to various friends of mine . . ." (Drake 1968, 91). From there on it was the all pervading "I" of the writer who sees by himself, "believes," "reasons," "holds," "questions," and so on. Even when Galileo promised to describe a method (discovered "by a pupil of mine" [ibid., 115]) of drawing the spots with complete accuracy, he easily slipped from the imperative mood into a discursive "I." Hence he began with "Direct the telescope . . ."

(ibid.) and after just three sentences he continued with, "In order to picture them accurately, I first describe . . . I find the exact place . . . I have drawn" (ibid.).

In order to establish the authority of an interpreter of natural phenomena, of the scientist who could offer the only true analogy (in this case, between the physics of the heaven and the physics of celestial bodies), Galileo resorted to many rhetorical strategies. He observed nature alone and related his own observations along with those of others in the first person. He created an image of himself as a persecuted individual, and used this image to suggest his moral superiority, for he claimed that the individual did not gain any advantage by breaking away from the community. Rather, setting himself against the "herd" had brought him more vexations than privileges:

And I, indeed, must be more cautious and circumspect than most other people in pronouncing upon anything new. As Your Excellency well knows, certain recent discoveries that depart from common and popular opinions have been noisily denied and impugned, obliging me to hide in silence every new idea of mine until I have more than proved it. (Ibid., 90)

Sometimes he acted as an instructor, teaching others the language of nature by telling them how to represent it. At other times he even appropriated nature's own voice to admonish nonauthorized observers that "Nature, deaf to our entreaties, will not alter or change the course of her effects" (ibid., 136). No wonder that toward the end of the third letter he arrived at the succinct formulation of the authority of the scientist alluded to earlier, "[F]or in the sciences the authority of thousands of opinions is not worth as much as one tiny spark of reason in an individual man" (ibid., 134). This individual, however, was somewhat depersonalized. His reading of natural phenomena was not subjective, neither was it arbitrary. Rather, he was completely tied to nature's own language, which he was obliged to learn before offering his interpretations. And the language, through God's providence, was accessible to anybody ready to acquire it:

[We] must recognize divine Providence, in that the means to such knowledge are very easy and may be speedily apprehended. Anyone is capable of procuring drawings made in distant places, and comparing them with those he has made himself on the same days. (Ibid., 119)

In Galileo's discourse the individual, detached from his community, but capable of creating exclusive conditions of enunciation, visibility, and manipulativity in which nature is decoded, was the source of authority. Making unobservables—the motion of the earth, inertial motion—major objects of explanation entailed a radical shift in the accepted rules of discourse and the organization of the intellectual map. The interpretation of visibles could not stop short in front of invisibles, whose nature remained unclear. Rather, it became necessary to subordinate the visible to the invisible so that the gap between them was seemingly effaced. In the course of this process, interpretation became seemingly superfluous, replaced by the pretence of the human mind to dominate the language of nature. The members of the new community were not interpreters, using their own language to "understand" nature, but rather scientists capable of speaking the language of nature.

REFERENCES

Biagioli, M. 1993. *Galileo Courtier: The Practice of Science in the Culture of Absolutism.* Chicago: University of Chicago Press.
Bourdieu, P. 1984. *Distinction: A Social Critique of the Judgment of Taste.* Translated by R. Nice. Cambridge, Mass.: Harvard University Press.
———. 1993. *The Field of Cultural Production: Essays on Art and Literature.* Edited by R. Johnson. New York: Columbia University Press.
Dear, P. 1987. "Jesuit Mathematical Science and the Reconstitution of Experience in the Early Seventeenth Century." *Studies in History and Philosophy of Science 18*: 133–75.
Drake, S., trans. 1957. "Letters on Sunspots." In *Discoveries and Opinions of Galileo.* Garden City: Doubleday, pp. 89–144.
Favaro, A., ed. 1968. *Le Opere di Galileo Galilei.* Vols. 1–21. Reprint. Florence: Olschki.
Feldhay, R. 1987. "Knowledge and Salvation in Jesuit Culture." *Science in Context 1*: 195–213.
———. 1995. *Galileo and the Church: Political Inquisition or Critical Dialogue?* New York: Cambridge University Press.
Foucault, M. 1972. "The Discourse on Language." Translated by R. Swyer. In *The Archeology of Knowledge.* New York: pp. 215–37.
———. 1977. "A Preface to Transgression." In D. F. Bouchard, ed., *Language, Counter-Memory, Practice: Selected Essays and Interviews.* Ithaca: Cornell University Press, p. 35.
Galilei, G. 1962. *Dialogue Concerning the Two Chief World Systems.* Translated by S. Drake. Berkeley and Los Angeles: University of California Press.
Koyré, A. 1978. *Galileo Studies.* Translated by J. Mepham. Sussex: Harvester Press.

Salmone, M., ed. 1979. *Ratio studiorum: l'ordinamento scolastico dei collegi dei Gesuiti*. Milan: Fettrinelli Economica.

Scheiner, C. 1968. "Tres epistolae de maculis solaribus," "De maculis solaribus et stellis circa Iovem errantibus accuratior disquisitio." In Favaro, vol. 5, pp. 20–70.

Schmitt, C.B. [1969] 1981. "Experience and Experiment: A Comparison of Zabarella's View with Galileo's in *De motu*." In *Studies in Renaissance Philosophy and Science*. (Originally published in *Studies in the Renaissance* 16: 80–138.) London: Variorum Reprints.

Comment: A New Way of Seeing Galileo's Sunspots (and New Ways to Talk Too)

Peter Machamer
Department of History and Philosophy of Science, University of Pittsburgh

R. Feldhay's essay is original, insightful and pellucid. Her focus on Galileo's rhetoric is an attempt in itself to break down the science-rhetoric, proof-persuasion dichotomy in much the same way that she sees Galileo to have done. By speaking of constituting scientific objects, constructing (disciplinary or field) boundaries, and by speaking about the authority of speakers, she sets a way of looking at Galileo that broadens out vision not only about him but about the nature of science (and, perhaps, about ourselves).

The explicit contrast of her approach is to all those who would see science as just collections of discoveries and proofs, and who would see the business of the historian and philosopher of science as documenting the discoveries and analyzing the progress of proof techniques. The rhetorical or Foucault-esque categories of discourse, practices, boundaries, and authority are meant to contrast with the older analytic terms such as explanations, problem solving, disciplines, backgrounds, predecessors, and progress or change.

New terms of analysis sometimes bring with them new ways of seeing, or, more accurately, new strategies for focusing attention on aspects that were differently selected, categorized, or labeled, or which had been ignored. In Feldhay's essay we are not asked to decide whether Galileo was the first modern mathematical physicist or the last Medieval, whether he was an Aristotelian or a Platonist, whether he relied mostly on experimental or mathematical reasoning, or whether he had a "modern" view of dynamics. This is all to the good. These categories of analysis, while they were important and effective in their time (from Duhem to Drake), have become moribund.

Less clear is that talk about Galileo's changing scientific discourse to include analogies between heavenly and earthly bodies brings more insight than the earlier talk about his breaking down the celestial-terrestrial distinction. It is less clear that talk about manipulativity being important in constituting scientific objects does more than speaking about Galileo's criteria for adequate knowledge and including as one criterion the ability to construct—real or imagined—mechanical models. Finally, most problematic, is that applying the category of authority to contrast the individualistic with the consensual illumines Galileo's epistemic challenges to Aristotelian and Church authorities more than talking about how, in the late sixteenth and early seventeenth centuries, the Renaissance anti-authoritarianism was a partial cause of the rise of a new neo-Protagorean individualism in all its epistemological, entrepreneurial, and governmental dimensions. Of course, talking about authority and power (of the group versus the individual), about breaking the boundaries or conventions as to what is legitimate, and about the constructivist epistemology of knowledge production does make the political paramount. In fact, it makes the political ubiquitous. The question then becomes whether anything can be contrasted with the political, and what, by virtue of this contrast, makes these specific political dimensions superlatively important. (And here one wishes that perhaps there were a little more C. W. Mills than Foucault as a model.)

Rather than develop these issues for the sake of contemplation and controversy, I propose to deal more specifically and directly with Feldhay's specific claims.

Let us start by looking at the end of her essay. Feldhay's Galileo in 1612–1613 seeks authority for his individual utterances so that he can rightly and legitimately claim to be a philosopher, and so make acceptable and authoritative pronouncements concerning the nature of things. According to Feldhay, Galileo had to establish the individual (as opposed to the group) as authoritative, and had to break down the boundaries of astronomy and natural philosophy, in which each practitioner had a distinct, well-proscribed job, in order to introduce a new scientific discourse in which a thing was "constituted as a scientific object by mathematical enunciation and by being given to observation . . . [by being] amenable to manipulation, even if only by analogy."

As mentioned, the claims concerning authority in the final section are confusing. Briefly, Feldhay argues that Galileo needed to make people credit the authority of the individual (which he did by the rhetorical devices of self-representation and the myth of being against the crowd) in order to justify the centrality of the analogy between the motion of the sun and the motion of heavy bodies on earth, that is, in order to establish and use the inertial principle to "prove" that the sun moves (and so, too, the earth).

I agree that the analogy between the terrestrial and celestial is central in Galileo's thought and his arguments during this period (see Machamer 1972). However, I think Galileo intended a conclusion far stronger than that stated in the analogies, though he often used terrestrial and celestial analogies to support the stronger case. he wanted his audience to conclude that only one realm of laws existed. He thought he showed in *Sidereus Nuncius* (1610; see Favaro 1968, vol. 3, sec. 1), before the "Letters on Sunspots" (1613; see ibid., vol. 4), that the laws of optics applied to the moon as well as to the earth (i.e., shadows on the mountains of the moon could be analyzed in the same way as the mountains of Bohemia), that the moon was earthlike, and that Jupiter and its "stars" were like a small solar system. In this regard he followed Tycho Brahe's lead (with his celestial comets and his comments on Copernicus) in bringing about the collapse of the celestial-terrestrial distinction in terms of the different kinds of laws that applied.

More specifically, like Brahe, Galileo did insist, as Feldhay often and rightly notes, that physical considerations must have their place in astronomical reasoning. However, he did so in 1610 before he achieved the title of philosopher, but, of course, not before he desired such a title. Even by 1610, he had, as W. Wallace (1984) has documented, studied with the Jesuits of the Collegio Romano to learn the philosophical ways of thinking and writing. But, surprisingly then, hardly any "neo-Aristotelian natural philosophical" terminology occurs in *Sidereus Nuncius*. At least in this way he did not intentionally attempt to demonstrate that he was a natural philosopher or show that he aspired to such authority. In fact, probably the success and popularity of *Sidereus Nuncius* invested him with authority. So, after that success, Galileo's status changed rapidly (as witnessed by his new job in the Medici court and the title of philosopher) and he, therefore, was in a position to influence what constituted the proper subject matter of astronomy. This, of course, leaves open the question of why, in 1610, *Sidereus Nuncius* became so popular and its author, who had done little before, so revered and important. But this is another question.

Having shown how Galileo became an authority figure, what the use of the inertial principle and the earth-heaven analogies have to do with the authority of the individual versus the consensual group remains unclear. Galileo's authority in the "Letters" derived from his prior success in his popular *Sidereus Nuncius*. Furthermore, referring to the first part of Feldhay's essay, Galileo's strong insistence on individual observation and individual reason, plus what Feldhay calls manipulability (i.e., something manipulable by an individual), seem to be constitutive of individual epistemic authority. Galileo, indeed, was one of the first (apart from the alchemists) to insist that knowledge was gained only by an individual interacting with nature by observation and reason. He did this without having developed any detailed individualistic epistemological theory. He was a poor philosopher in many ways. But the individualism, preached by Galileo in elementary Latin, and later in the Italian vernacular, along with novelties to be seen in the sky, were heady and popular messages.

Let us turn to another point that occurs in Feldhay's title, later developed in her discussion about manipulation and the "experiment" with the bitumen on the red hot iron plate. While the concept of *praxis* is dear to many, and modes of production that actually produce material goods are supposed to be cherished whenever they can be found, I do not think Galileo's experiment is indicative of, or about, manipulation. If his point involves manipulation it does so in a complex and strange way. The mentioned experiment of the hot plate demonstrates much about the nature of Galilean experiments, mechanical models, and his whole view of what is intelligible in science.

Briefly, in the space available, I will lay out the claims without the supporting evidence. From the time of *De Motu* onwards (c. 1590; see Favaro 1968, vol. 1, pp. 251–419) Galileo held that the way to understand anything was to show how it worked in a mechanical way. This probably derived from his practical training in the mechanical arts, his so-called artisan-engineering background. This was also a reasonably new Renaissance enterprise, and was individualistic in being developed outside of traditional guilds or schools. It was a practical do-it-yourself, apprentice training set of activities. It was practical in that the practitioners designed instruments, built fortifications, and constructed machines. One could understand how these machines worked by constructing them, and seeing what they did. For big projects one built simplified models, or found everyday experiences that exemplified the relevant principles.

In this sense I agree with Feldhay that Galileo added to the traditional criteria of mathematical description (from the mixed sciences) and observation (from astronomy) for constructing scientific objects (as she would say) or for having adequate explanations of the phenomena observed (as I would say). Intelligibility for Galileo required having a mechanical model of the phenomenon to be understood, and mechanical models such as the balance of the inclined plane were manipulable. For motion problems this meant the model of intelligibility was the Archimedian simple machines: first the balance, and, then later, the inclined plane. In the theory of the tides the model was a constructed wheel with an embedded tube of water that could be rotated. For the moon's shining a puddle of water was analogized. The bitumen on the hot pan creating clouds of smoke falls into this class of models. These mechanical models were a necessary part of the "proof" or criteria of adequacy for determining a valid explanation.

Incidentally, I do not find talk about constructing new scientific objects in this context useful since Galileo uses this ploy in talking about the motion of bodies which is clearly a known scientific phenomenon, as contrasted with what he thinks are new sciences, never before seen, that is, his sciences of materials and local motion. He also uses the model criterion for his account of the tides, which had been an object of study for centuries.

However, constructing or finding models is a criterions for understanding the physical workings of a given phenomenon. That most such models are manipulable in some sense is true, but Galileo never emphasizes the manipulability as important, except insofar as they can be constructed (at least in principle). I think this also relates to Feldhay's point about the necessity of imagining motions in that what is mechanical in this sense can be drawn or reproduced in a picture or recipe book. Such things can be seen or made by every*one*, every individual.

These are minor points of disagreement with Feldhay. For most everything I agree with her new and important insights, though I might rewrite the way in which her claims are phrased. Particularly I agree that Galileo's "Letters on Sunspots" were not wholly or even primarily concerned with establishing merely the facts that the spots were contiguous with the surface of the sun and were generated and dissolved as they passed across the body of the sun. He was concerned with Copernicanism, and so, as she well points out, with the rotation of the sun (and so of the earth), and with its corruptible, maculate (as opposed to incorruptible, immaculate) nature.

Feldhay nicely sets up Galileo's problem of trying to make intelligible the physical motion of the sunspots. He is just better at observation and mathematical interpretation than Scheiner, and shows this by his mixed science and perspectival arguments about foreshortening. As she cleverly shows, this leaves a gap for Galileo which he needs to fill in order to account physically for the motion of things on the sun. He sees that rotation fills the gap, but then he must make plausible that the sun and its ambient rotate. So he looks to, or develops the idea of, "indifferent" or inertial motion. This relies basically on a sufficient reason argument, namely, once in motion always in motion for what should stop it. He applies the argument counterfactually to a ship traveling across the seas, and then turns it to the sun. Since the spots *seem to be* rotating, the sun *is* rotating, and it *will continue* to rotate. But the brilliance in Galileo's move, and in Feldhay's analysis, comes by bringing in the analogy to clouds. Clouds move along with the earth's ambient—the air. So too the spots move along with the sun's rotation, the unstated conclusion being that the earth rotates, too.

Feldhay's discussion of the boundaries of astronomy versus natural philosophy is most helpful. Undoubtedly, Galileo wanted to be thought of as a philosophical astronomer, and so was empowered to deal with physical substances and their causes. As I noted he dealt with such physical issues earlier in his *Sidereus Nuncius*, and so continued a path opened by Brahe when he tried to "physicalize" Copernicus. Such physical and causal concerns were also true in his study of motion. After his practical training at the hands of Ricci and Cigioli, he found his new science of local motion—science by way of machines.

But this was not philosophical in any accepted sense. He needed to learn to talk about the causes in ways in which traditional natural philosophers could understand. To this end, as Wallace has shown, Galileo read philosophy and tried to adapt his insights into a philosophical framework and deal with philosophical problems. With this background he wrote, pleading for a job at the Tuscan court, to Cosimo's secretary, B. Vinta, "I desire that in addition to the title mathematician His Highness will annex that of "philosopher"; for I may claim to have studied more years in philosophy than months in pure mathematics" (Favaro 1968, vol. 10, 348; my translation).

Finally I agree with a general point that Feldhay makes toward the beginning of her essay: that the separation of science from rhetoric is a strange distinction. However, I find that the distinction is held by as many rhetoricians as it is by positivisticly leaning historians and philosophers of science. I think Feldhay and I agree that rhetoric should not be limited to the study of traditional tropes or devices, nor science or philosophy only to demonstrative proofs. Certainly, scientists have never so limited themselves, no matter what some few of them may have professed. This is better appreciated when studying how what counts as a proof has changed over time. On the one hand, what is properly called "demonstration," and what it means to demonstrate, has changed dramatically at least three or four times. On the other, argument by analogy always has been studied and sometimes, as in many fourteenth- and fifteenth-century thinkers, it has been included in the canon of logic. Again, despite Plato, there is no antithesis between proof and persuasion, for proof, however construed, has always been taken to be the best or most legitimate way in which to persuade; it has been called persuasion by appeal to reason.

Science of course has a rhetoric. This means looking at the texts of the scientists and the way that their words are used. To do this one must understand the context, institutions, and practices which gave those words their meanings. This is not the meaning of formal syntax and model theoretic semantics. These are of little or no help to the historian or philosopher of science. The meaning we must seek lies in the way people interact with the world, and how they build images or models to understand and interact with that world, to make that world intelligible. We might even say it is our job as historians of science to figure out how they imagined the world.

REFERENCES

Favaro, A., ed. 1968. *Le Opere di Galileo Galilei.* Vols. 1–21. Reprint. Florence: Olschki.

Machamer, P.K. 1972. "Feyerabend and Galileo." *Studies in History and Philosophy of Science* 4: 1–46.

Wallace, W. 1984. *Galileo and His Sources.* Princeton: Princeton University Press.

7

Rhetoric and the Cold Fusion Controversy

From the Chemists' Woodstock to the Physicists' Altamont

Trevor J. Pinch

Department of Science and Technology Studies, Cornell University

> I have as witnesses most excellent men and noble doctors . . . all acknowl-
> edged that the instrument deceived. And Galileo became silent . . . dejected,
> he took his leave . . . And he gave no thanks for the favors and the many
> thoughts, because, full of himself, he hawked a fable . . . Thus the wretched
> Galileo left Bologna with his spyglass . . . .
> —Martin Horky's account to Johannes Kepler of Galileo's visit to Bologna

Rhetoric is the art of persuasion. More precisely, it may be defined
as the systematic attempt by one or more social actors to *persuade*
other social actors to do something or that something is the case. Rhetoric can be used, for example, by a politician to argue the merits of
being a democrat, by a salesperson to promote the bargain status of
some goods for sale, or by a scientist to establish the existence or non-
existence of a scientific entity. The means of persuasion can include
written texts, such as newspaper articles, sales brochures, or scientific
papers; speeches, such as political speeches, sales patter, or conference
presentations; visual displays, such as charts of unemployment, the
"flash" (display) of goods of a market trader, or the overhead graphs
of a scientist; and demonstrations, the kissing of babies on a politician's "walk about," the showing off of goods by a market trader
(known as "doing the dem"), or the demonstration of the working of
a particular experiment.

Discussions of rhetoric in the context of science encounter the familiar difficulty that science, as the exemplification of valid reasoning, is seen as antithetical to rhetoric. This difficulty stems from the history of rhetoric since Aristotle where the tendency has been to separate valid reasoning, such as logical reasoning, from rhetoric. The difficulty is compounded by scientists' own accounting practices whereby they routinely invoke social factors only to explain error (Gilbert and Mulkay 1984). Thus in the epigraph, Horky refers to what he takes to be Galileo's mistaken observations of the moons of Jupiter as "hawking" a fable. In regards to science it almost seems to be a contradiction in terms to talk about "hawking a successful scientific theory." As Latour (1987) notes, scientists maintain that they are persuaded of the truth of matters only when all sources of persuasion have vanished.

The study of scientific rhetoric has thus become denigrated. This is further reflected in common parlance where one finds reference to "mere rhetoric" or "rhetorical flourishes" as it these were epiphenomenal to the main business at hand.

In that science is about valid reasoning it seems that there would be very little purchase for an analysis of rhetoric. That rhetorical analyses of science are increasingly prevalent is as much an indication of our changed understanding of science as it is of a newfound confidence in the claims of rhetorical analysis.

The field of the sociology of scientific knowledge has played its part in our changed understanding of science. Sociologists of science have shown with many case studies how the very content and technical practices of even the hardest of sciences can be viewed in terms of processes of social construction (e.g., Latour and Woolgar 1979, Collins 1985, Pickering 1984, Pinch 1986, Latour 1987, Lynch 1986, Shapin and Schaffer 1985). Symmetry, whereby what is taken to be scientific truth and error is analyzed in the same way, has become the watchword for these new studies (Bloor 1976). If scientific outcomes are shaped by social factors, why should not rhetoric enter the picture as yet another factor whereby the contingency of scientific truth is established? Indeed scholars within the sociology and history of science have specifically pointed to the role of rhetoric in science (e.g., Dear 1985, 1991; Shapin and Schaffer 1985). Students of rhetorical analysis and literary criticism have also recently turned to science and have shown how, from a variety of perspectives, it can be analyzed with their standard tools of rhetorical analysis (e.g., McCloskey 1985, Bazerman 1988, Myers 1990, Gross 1990, Locke 1992).

With science no longer the special case it once was, the onus has switched toward understanding how rhetoric in general works. The rhetorical project would, in its most ambitious form, search for similarities in rhetorical structures in a range of discursive domains including science. If one takes a broad definition of rhetoric—how social actors persuade other social actors—then a large number of different domains are to be examined.

In this essay I look at two such domains which on the face of it seem very different: science and selling. Indeed for some people they are polar opposites. Science is often taken to be the exemplification of all that is good and noble in human endeavor—the pursuit of disinterested, objective, value-free knowledge—whereas selling is often taken to be about personal interests and pecuniary gains—in short, about making money. We *revere* our scientists, and *sneer* at our used-car sales representatives.

However, we can learn useful lessons from the study of selling. Refreshingly, it is an area where one does not have to keep looking over one's shoulder for philosophers wielding the cudgel of valid reasoning. Indeed, one can say that the problem one faces in studying selling is to find anything which looks like valid reasoning at all! Selling is the area par excellence where persuasion is to the fore. By comparing such a seemingly different domain to science, I hope to lay to rest forever the argument that science is a special case for which the tools of rhetorical analysis are less applicable. If the same rhetorical devices can be observed working on the back streets of Leeds Market as in the hallowed halls of academy then the scope of a rhetoric of science would seem to be very wide.

Having pointed to obvious differences in the status we usually grant the activities of science and selling, and before discussing the details of the sorts of rhetoric to be found in both areas, let us note that some congruences are to be found. Scientists may refer to their attempts to get funding for large experiments as "selling." For instance, one solar-neutrino scientist, J. Bahcall, recalls how, before a particular meeting in which an influential scientist's support was needed for a project, he and his collaborator carefully prepared in advance in order to try to sell the project to the scientist:

Ray's [Davis] opinion was well we had to sell Maurice [Goldhaber] or else we wouldn't get the experiment. Ray's opinion was that this nuclear physics trick would be something that Maurice would be turned on about, because he was himself a very bright nuclear theorist. . . . He would think that that's a cute idea and get turned on about it. So we decided to sell him the experiment based mainly on this trick. Then more or less that worked. (Quoted in Pinch 1986, 86)

Clearly in that some scientific experiments and facilities cost large amounts of money and scientists have to persuade hard-pressed financial backers to provide the necessary support, scientists can be seen as being involved in a form of selling. This is transparent when seeking money for a large facility that requires the support of politicians. In such cases the political lobbying and the need to persuade the public at large about the merits of the case are obvious. Indeed, the current effort (as of 1991) to persuade the U.S. Congress to support the supercollider can be said to be no more than encyclopedia selling writ large. The encyclopedia salesperson's task is to persuade the customers why they should part with large sums of money for something of which the main benefit is knowledge—so too for the high-energy physicist seeking support for the supercollider. (Of course, the types of persuasive resources employed in both cases differ. Buying an encyclopedia does not produce jobs, spin-off, or votes—what is good for your family may not be good for Texas as a whole.) Even away from these large-scale projects, as already pointed out, other occasions in science require that people be sold on a particular idea. Although no monetary transactions may be involved, that scientists can think of this activity as being like selling indicates the likely fruits which such a comparison might bring.

Rhetoric in Selling

The work on selling I have conducted commenced with the study of a group of sellers in Britain known as "pitchers" (Pinch and Clark 1986; Clark and Pinch 1988, 1992, 1993, 1995). These are salespeople found largely on markets who employ an extended oral sales spiel or patter to sell to a large crowd. Usually pitchers have loud voices or use a microphone. Their sales routines are characterized by much humor and interaction. Sales are made en masse at a fixed point in the routine. The pitchers are the original "patter merchants."

Pitching is the traditional way to sell goods on British markets; in Victorian times most goods on markets were sold by these means. Today usually only one pitcher is found on any market, although special Sunday markets and tourist sites (such as Covent Garden in London) attract more. C. Clark and I have observed pitchers at work in other European countries such as France and the Netherlands and observations have also been made of pitchers working at country fairs in the U.S. (Sherry 1988).

By the use of audio and visual recordings, Clark and I have studied in detail the sales rhetoric pitchers used. We have found that, although a variety of standard routines are used, these all embody similar devices for describing the goods, for making their bargain status apparent, for rendering prospective purchasers under obligations to buy, and for obtaining mass purchases (for details see Pinch and Clark 1986, and Clark and Pinch 1995). Perhaps the most common device is the use of a rhetorical form known as a contrast. Pitchers mainly contrast a high worth and a low price for the goods. The contrast is embedded in a descending list of possible prices which might be paid for the goods:

All the kids love Jedi an' Star Wars. 'ere ah don't want four pound, two pound, one fifty, an' not even one twenty. Ah've got two lads've me own, one's five months an' one is six on Wednesday. Ah know what kids go for. Yuh can 'ave anything on the board I call 'em stocking fillers. Whatever yuh like (clap) a pound a piece. (Quoted in Pinch and Clark 1986, 177).

In our most recent research, Clark and I study more orthodox areas of selling such as field selling, telephone selling and door-to-door selling (see Clark et al. 1994, Mulkay et al. 1993). The assumption behind our work is that goods do not sell themselves and that the skillful seller is at the same time a skillful rhetorician who has to persuade the customer.

How Effective Is Rhetoric?

Clark's and my choice of methods and research site stems in part from a major difficulty facing any rhetorical analysis: the problem of ascertaining whether rhetoric is actually effective. In short how do we know that presenting goods for sale one way as opposed to another is more effective in terms of the desired persuasion? Similarly, in thinking about rhetoric in science, how do we know that presenting an argument one way is more effective than another for recipients? We have found two main ways to address this issue.

1. We look at locations where we can monitor audience responses

directly. An obvious example of such research is M. Atkinson's (1984) work on the effectiveness of political rhetoric. Atkinson recorded political speeches and monitored the audience's applause. Such speeches are frequently interrupted by applause. Atkinson studies the specific rhetorical devices which immediately precede the applause. Similarly, in our work on market traders, we study some of the desired response directly by videotaping the actual point of sale.

2. We search for similarities of devices within and between different domains and different audiences. The same devices that occur repeatedly gives us some grounds for assuming that orators find such devices effective. That is, the assumption here is that orators attend to the effectiveness of their rhetoric and that good orators will repeatedly use the same devices. Thus in our work on market traders we have documented the same devices being used again and again in a variety of sales locales, by a variety of pitchers, and for a variety of products.

As a supplement to these two main methods, we can study how recipients view the effectiveness of a particular piece of rhetoric by interviewing them about, say, a speech to which they have listened or a text they have read. Audiences can be quizzed as to which bits of a speech or sales pitch they particularly remember or what was particularly effective for them. (This is similar to what happens after the televised U.S. presidential debates when the major TV news networks survey audiences for their immediate responses.) Such a method is of course heaped with difficulties as it depends upon a response given in a new discursive context—an interview—being an accurate assessment of what happened in another discursive context—the original rhetorical situation. It also ignores the subtlety of the rhetorical process whereby often people are convinced by devices which they might not recognize as such.

Another supplementary way of tackling the problem of what effect rhetoric has is to study the dissemination of the speech or text in yet further texts. For instance, political speeches are often reported in newspapers and similarly talks at scientific conferences provide commentary for wider dissemination, whether in popular books or popular science magazines. In other words, some pieces of a text (perhaps rhetorically the most effective elements) will be widely quoted and reproduced in yet other texts. Thus, for example, Atkinson finds that the parts of political speeches most widely quoted, say in a subsequent news clip, take the exemplary rhetorical form of a contrast. Indeed the parts of political speeches we tend to remember also take this exemplary form, as in the classic Kennedy speech, "Ask not what your country can do for you, but what you can do for your country."

None of these considerations should be taken as detracting from the subtlety of the rhetorical process. Thus I do not suggest a mechanical relationship of rhetorical device leading to rhetorical effect. The production of meaning depends upon a speaker and an audience coconstructing the meaning of the text. Audiences are not rhetorical dopes and may themselves be monitoring for the use of any blatant rhetorical devices—similarly the skillful orator is aware of the recipients monitoring his or her talk for such devices. Also the employment of any rhetorical device will undoubtedly be sensitive to the context of use including what has been said before, what is to come next, and the overall context.

Toward a Comparative Rhetoric

One way of thinking about this project is to list the different types of persuasive work which actors routinely need to do. Then, having outlined typical sorts of situation where rhetoric might be employed, one could compare the types of rhetorical device found in different domains. For example, most actors have to draw boundaries around their area—what is the rhetoric of inclusion and exclusion? They have to build rapport with their audiences—how is this done in different circumstances? They have to accomplish similarity and difference; some things are constructed as being similar to what they do and some as different. Actors will face particular problematic junctures in their rhetoric, for example, having to give bad news, managing a problematic action or going against an audience's expectations. They may have to praise some groups and castigate others, or deal with self-praise (such as in Nobel prize speeches [Mulkay 1984]). The list is potentially endless, but the need for comparison is essential.

It's Not an A, It's Not a B, It's a C. As a starting point, and to give an indication of the ambitions I have for this project, let us examine briefly how one frequently encountered sort of persuasive task can be handled similarly in science and in selling. This is the task of convincing others of the merits of something about which they know a little. The thing of interest, in the case of science, might be a set of ideas or a technique; in the case of selling, it might be a product or a price.

My first example is taken from a historic conference set up to discuss a rising new specialty in sociology called ethnomethodology. As well as practitioners of ethnomethodology, the audience consisted of informed neutrals and skeptics. The leading doyen of ethnomethodology, H. Garfinkel, opened the conference. Rather than state what ethnomethodology was, Garfinkel chose to say what ethnomethodology was not:

> To begin with our work is not a mysterious enterprise and we are not peddling remedies. You may not believe it now, but it is not a cult. It is not an enterprise directed to the solution of whatever it is that we think ails sociology. Last of all, is it any such pretentious thing! It is not declaring war and, even while it disturbs you, it is not even a position although that is probably the most head-scratching thing of them all. Well if it is not a position that we are providing, then what are we doing here? (Quoted in Hill and Crittenden 1968, 3–4)

This way of introducing something, by listing a number of things which it is *not* and then contrasting the list with what it *is*, is evident also in the following extract. Here a health economist makes a presentation to a conference of clinicians. She introduces the new concept of "management budgeting":

> This is very, very important. Management budgeting is *not* a costing system, it's not a glorified cost accountancy system. It's about management, managing resources. (See Ashmore et al. 1989, 127)

We find that this strategy is also evident in selling. It is, for instance, the predominant way that pitchers introduce the price of an item. As is evident in the pitcher's extract in the second section, the price is introduced by listing a series of prices which the item will *not* cost which is then contrasted with the final price. A further example comes from my observations of double-glazing sales representatives. One salesperson introduces his product by handing over to the customer for inspection a piece of a window which appears to be double glazed. It transpires that this sample comes from the competition and, having pointed to several faults with the sample, he makes a contrast with his own company's product which is then handed to the customer. In this example the listing and contrast device occurs not only in speech but by way of demonstration.

One last example of this type of rhetoric comes again from the area of health economics. In a series of articles published in the *British Medical Journal* two leading health economists argue for health economics to play a larger role in medical decision making. In the first of the series, introducing what health economics is, the two economists offer a series of definitions of what health economics is not, or as they put it, "common misconceptions about economics." They list a number of such misconceptions: that economics is about money, costs, cutting costs, and accountancy. In other words, by attending first to what doctors might have taken health economics to be about, the stage is set for saying what health economics actually is (ibid.).

It seems that in all these examples the listing of what a set of ideas-product-price might be thought to be like, but is *not*, is a felicitous way of introducing what the ideas-product-price *is*. The type of rhetorical device at work is a particular combination of lists and contrasts. The use of lists and contrasts is widespread in both the political and selling domains (Atkinson 1984, Pinch and Clark 1986). By putting peoples' misconceptions about something at the heart of the contrast, sensitivity is displayed toward the audience-reader's own knowledge. Rather than introducing the new set of ideas, price, or product "out of the blue," this method takes account of the reader-audience's own understandings (or misunderstandings) and in this sense can be seen, at the same time, to build rapport with the reader-audience. It builds solidarity between speaker-author and the audience-reader, which is crucial to the persuasive task at hand.

Rhetoric and the Cold Fusion Controversy

I now turn to some examples of rhetoric from the cold fusion controversy. My intention is not to analyze the history of this episode in other than the most cursory way. Other articles have examined the controversy from the viewpoint of a number of issues in the sociology of science and science communication (see, e.g., Collins and Pinch 1993, Lewenstein 1992, and Gieryn 1992). The essentials of the story centers on the claims of two chemists, M. Fleischmann and S. Pons, to have observed nuclear fusion in a test tube. The purported discovery, announced in March 1989 at an infamous press conference at the University of Utah, attracted much attention. Although it was difficult to understand how Pons and Fleischmann could be seeing such large heat excesses if the reactions were truly nuclear in origin (because, according to standard nuclear theory, the neutrons produced should have been more than enough to kill both experimenters), the work seemed of sufficient promise for many scientists to get involved. Furthermore, the announcement, within a week, by another group at nearby Brigham Young University led by S. Jones, that they also had found evidence for cold fusion, reinforced the feeling that something new and dramatic was in the air.

On 12 April 1989 Pons met a rapturous reception when he lectured to 7,000 chemists at the American Chemical Society (ACS) meeting in Dallas. It was, as commentators noted, the Woodstock of cold fusion. However, less than three weeks later at the American Physical Society (APS) meeting at Baltimore on May 1, where a number of papers were presented highly critical of Pons and Fleischmann, Woodstock turned to Altamont (site of the Rolling Stones' concert in the 1960s at which one of the audience members was stabbed to death while the band played on, thus ending the famous "summer of love").

One of the advantages of focusing on the cold fusion controversy is that much of the action happened in public forums—press conferences, scientific conferences, media interviews, and so on. This means that we have unparalleled material to examine in terms of rhetoric (much of this material is located at Cornell University in the Cornell University Archive for the History of Cold Fusion). My initial focus has been upon video recordings of presentations at the two conferences.

Humor and Rhetoric

The conference presentations of scientists are expected to differ from either the speeches of politicians which are constantly interrupted by applause or the sales patter of pitchers which lead to many sales. Certainly this is true for most conference presentations by scientists where the audience typically applauds at the end and perhaps laughs at one standard joke.[1] Seemingly such public presentations pay attention to the formal norms of science (Merton 1942). The claims made tend to be supported by argument and evidence and particularistic or ad hominem claims are avoided; the tenor of such talks is usually that of the disinterested skeptic. Key presentations, however, at both APS and ACS conferences were punctuated by laughter and sometimes laughter with accompanying applause:

There was no trouble finding which hall the action was due to take place in as the noise from the developing crowd identified it at two floors and fifty meters range. Well over a thousand people were already there; media took care of the right hand side of the hall and the front five rows were roped off for speakers and officials of the APS. This was unlike any physics convention I had experienced in twenty years; it was more reminiscent of a political convention. (Close 1990, 212)

These then were no ordinary scientific meetings. A lot was at stake and the audience size and media interest were unprecedented. Indeed the cold fusion session at the ACS meeting had to be hastily transferred to a nearby basketball stadium to accommodate the crowd of chemists.

Unlike with most contexts in science, we have here an opportunity to study visible manifestations of the audience's response. The first thing to note is that the audience is not laughing and clapping in a mechanistic way in response to particular humorous triggers prepared by the speaker, such as during the performance of a stand-up comic. During APS President J. Krumhansl's speech to open the cold fusion session, apparently not written humorously, the audience constructs it as such:

The American Physical Society has as its purpose the advancement and diffusion of the knowledge of physics. As a part of science in an overall sense we do our part through open meetings and refereed scientific publications [Laughter from audience].

The laughter is produced by the contrasting image invoked by the phrase "refereed scientific publications" followed by a lengthy pause at which the audience took as a cue that laughter was appropriate. Most of the audience is aware of the allegations that Pons and Fleischmann have breached scientific ethics by going public before publishing their results in a refereed scientific publication. This example demonstrates the subtlety of the rhetorical process. The audience, with its laughter, constructs a particular meaning of the text which the author might not have intended. Of course, this does not detract from the rhetorical effect of the text; it does, however, further caution us against treating rhetoric in too mechanistic a manner, for, again, audiences are not rhetorical dopes. The process of constructing the rhetoric is interactive and the rhetorical effect is coproduced by the speaker and the audience.

What Does Humor Do? Before examining in detail some more of the instances which produced laughter, let us consider humor in general. Most theories of humor concur on the basic structural form that produces humor (Mulkay 1986). Two different frames or perspectives are brought into unusual conjunction and the result seems to be humorous, not to be confused with "laughter." (Humor accompanies many sorts of activities, but may not always result in hearable laughter. Formal joke telling is only one subset of humor. Much humor occurs spontaneously without a formal joke being told.) Humor therefore is about difference; it is about the unexpected bringing together of two different images or scenarios. For instance, the APS president unwittingly brings together the scenario of the careful refereeing of publications in physics with Pons and Fleischmann's supposed failure to have their claims refereed.

Why was humor so evident at these meetings? Here are at least two reasons. First, and most obvious, humor builds rapport and solidarity. Members of an audience laugh together in a similar way that they clap together. Even if someone does not see the point of the humor, that person experiences social pressure to laugh along with everyone else. Members of an audience constantly monitor each other for such public displays. Because laughing is a public response shared by members of the audience, it builds rapport and solidarity among the members.

A second reason for humor on such occasions is more subtle and reflective of the serious work that accompanies humor. As Mulkay (1986) argues, humor is a vehicle that allows serious points to be made, but the serious points are not accountable in the usual way because they have been marked by humor. For example, if I say in a humorous manner, "It is refreshing to study selling because one does not have to keep looking over one's shoulder for philosophers wielding the cudgel of valid reasoning," I am pointing to philosophers' using a certain form of reasoning against sociologists, but I am escaping from the consequences of making such an allegation directly. I only allude to it by trying to make the remarks humorous, but at the same time I am still saying them and for people to laugh at the remark (assuming they did find it funny) they must recognize in some senses the validity of what I am saying. In other words, humor can be a vehicle to make serious points *without* seeming to be serious and without carrying the normal responsibilities of being accountable for what we say.

I suggest that the instances of humor we will analyze in the cold fusion controversy are accompanied by deadly serious work. Perhaps the most important point about the serious work is that many of the things which generate humor seem to breach the normative considerations of scientific presentations. In other words, the humor allows particularistic claims to be made, for commonsense metaphors to be introduced, for ad hominem arguments to be made, and for unwarranted speculation to be put forward.[2]

Before examining material from the two conferences in more detail, and in keeping with the spirit of the comparative project on rhetoric, I will first give examples of how humor plays a part in the rhetoric of selling.

Humor in Selling

Humor is routinely used in selling to build rapport and solidarity. A typical pitching routine contains many jokes and humorous moments:

Yer get the Cindy Boy and the Cindy Girl (holds up two dolls riding a tricycle together) and the Baby (puts baby in back of tricycle). And the only reason the babies are here is we put 'er (points to Cindy Girl) in the same box as bloody Action Man (audience laughs).[3]

(Much humor, as in this case, centers around sexual insinuations.)

Market pitching involves a number of problematic actions; these are invariably managed through humor (Clark and Pinch 1988). In one example, pitchers want the crowd to move in closer together in order to elicit mass sales—that is, they must address the crowd as a crowd rather than as a set of individuals. The pitchers use humor to entice the crowd to move forward:

Now listen don't, don't, don't, don't get eager. Would you like to come forward for them. There's a new move, that's the idea, there's a new move that if yer put yer left foot forward, and then yer right, your body will follow. It's called walkin'. That's it. Lady over there. Come forward, that's the idea luv, yuh know I've been vaccinated.

Another example is the extended use of humor during a field sale (where a sales "rep" visits a customer in order to sell products). In this case the customer turns down the rep's sales proposals. In order that this should not create animosity, the customer often uses humor to reject the proposal. The serious work of saving "face" (Goffman 1974) in interaction is managed through humor (Mulkay, et al. 1993).

The Chemists' Little Science versus the Physicists' Big Science

One of the notable features of the cold fusion debate has been the claims of chemists to achieve a small science route to cold fusion, which is diametrically opposed to the big science, big bucks route of hot fusion. This difference in approach was at the heart of two of the occasions for sustained humor at ACS. The first extract to examine comes from the introduction to the cold fusion session by ACS president, C. F. Callis. Callis begins in a fairly straightforward manner, giving a brief history of hot fusion research. After clarifying how limited the achievements have been, especially in view of the large investment, he simply says, "Now it appears that chemists may have come to the rescue." This is greeted by enthusiastic laughter and applause:

Starting with Project Sherwood in the early 1950s the Federal Government and other countries have funded research that was and is hoped will lead to a method of initiating, sustaining and controlling the fusion of deuterium and tritium nuclei. Many billion dollars have been invested in attempting to reach a net energy gain through magnetic or inertial, that is laser or particle beam confinement, heating and compression of fusion fuel. This research generally involves temperatures of hundreds of millions of degrees, and in some instances comparable pressures. While much has been learned about plasma physics, and while much progress has been made, the goal has remained elusive, and the large complicated machines that are involved appear to be too expensive and too inefficient to lead to practical power. Now it appears that chemists may have come to the rescue. [Audience laughter and applause].

The humor in this case works by contrasting the scenario of the vast and largely unsuccessful efforts of physicists with the more modest efforts of chemists which promise much success. Clayton's remarks are also humorous because they conjure up a reversal in the normal hierarchy of the sciences whereby chemistry is usually seen as being subservient to physics. Now chemists have come to the rescue of the physicists. The same sort of humor occurs during Pons's talk. In a famed incident, Pons calls, in a deadpan voice in the middle of his presentation, for the next slide. It is a picture of his cold fusion cell. Pons announces:

Is a photograph of the U-1 Utah Tokomak. [Laughter and applause.]

This comment, of course, produces the same contrast between the low-tech chemists' approach and the big science approach of the physicists with their Tokomaks. More laughter follows as Pons remarks:

Early on we didn't have much money so we chose "Rubbermaid" here [laughter] which worked out quite well for the water bath.

Again this plays on the implied difference between the little science approach of the chemists and the big science approach of the physicists. The use of humor to emphasize these differences builds solidarity and rapport throughout the audience of chemists. Humor also facilitates the building of solidarity without transgressing normative considerations such as would be involved in making particularistic claims that one whole disciplinary group, chemists, are better than another, physicists. If Pons or Callis said directly, in a way that did not invite humor, something to the effect that "we chemists are better than the physicists because we are doing this by low tech means, using cheap ordinary materials," they would run the risk of sounding heavy-handed. Humor allows the normative considerations to be breached without sanction to the speaker.

The effectiveness of these pieces of rhetoric, apart from the direct laughter and applause produced in the audience, is further evidenced by the fact that the Pons incident with the slide has been cited in numerous accounts of the ACS meeting (e.g., Close 1990, 148).

The importance of the audience for generating the rhetorical effect can be seen by the contrasting reception accorded S. Jones (the head of the Brigham Young cold fusion group) when he appeared at the APS meeting. Jones tried the almost identical formulation to Pons of contrasting a picture of his own rather modest cold fusion cell with a photograph of a hot fusion facility (the JET at Culham). In showing the slide of his own apparatus Jones (paralleling Pons) had even mentioned the pennies that were used for shielding. In a deadpan voice Jones announces:

This setup [his own] is quite simple compared to this one [JET]. [Isolated laughter].

Although the contrasting visual images produces a few guffaws, it is not the mass laughter and applause that greeted Pons at the ACS meeting. For this audience of physicists, at this time, the joke was not taken as being at the physicists' expense.

The Rhetoric of Numbers—Little and Large

The relationship between the two Utah cold fusion groups has been tense. As well as a unsavory priority dispute, allegations have been made by both groups that one has plagiarized the other's ideas. Furthermore, the findings of the two groups have been different. The Brigham Young group has never claimed to find excess heat and has found neutrons at orders of magnitude less than those claimed by Pons and

Fleischmann. Throughout the cold fusion debate Jones at Brigham Young has been anxious to distance himself from Pons and Fleischmann's findings. Here, I examine one rhetorical move, accomplished through humor, that Jones has repeatedly used.

Essentially, Jones uses a particular metaphor to clarify the differences between what his group claims and what Pons and Fleischmann claim. The metaphor is the difference between a dollar bill and the U.S. budget deficit. The amount of excess energy Jones claims to observe is presented as being equivalent to a dollar bill, while the excess energy claimed by Pons and Fleischmann is said to be equivalent to the whole U.S. national debt:

> The ratio between our results and the Pons-Fleischmann result is the same roughly as the ratio between a dollar bill and the national debt. [Laughter and sporadic applause.]

This sample is taken from his talk at the APS meeting in Baltimore. He has used this metaphor to some effect on other occasions, such as when talking to the press. The following extract comes from the MacNeil-Lehrer News Hour of 28 April 1989. Note how in this context Jones embellishes the metaphor by using a real dollar bill accompanied by a number of gestures with his hands:

> If we let our results represent [takes dollar bill from wallet] one dollar, OK. Our heat output, then their results, you know [gesticulating with right hand to trace sharply rising upwards curve] would be ten trillion times as, would easily be enough to pay off the national debt [lifts up dollar bill in right hand and then lowers bill], you see that's the ratio [moving right hand] of the results, the national debt [gesticulating with the right hand] to a single dollar roughly.

The significance of metaphors as rhetorical devices is well known within rhetorical research (e.g., Gross 1990). In this case the metaphor does the work of translating a very large number expressing a difference into a numerical difference which is immediately familiar to the audience. We may not know exactly what the budget deficit is but we know it is huge compared to a dollar.

Again, as with the other pieces of rhetoric documented in the cold fusion debate, Jones's dollar bill metaphor has been widely quoted in the secondary literature covering the debate (e.g., Close 1990).

Koonin's negative talk also puts flesh on numbers to illustrate differences in size; in this case, a very small number is described. In the following extract Koonin illustrates the small likelihood, according to standard nuclear theory, of fusion occurring in Pons and Fleischmann's cell:

Nevertheless all these things [the probability of a particular fusion reaction occurring] are really small. How small in ten to the minus sixty-four per second for the dd reaction? Well its about one fusion in a solar mass of deuterium a year. [Laughter and sounds like one hand clapping.]

Part of the effect is achieved by the cleverness of the choice of metaphor. By extrapolating Pons and Fleischmann's test-tube fusion results to a body the size of the sun, Koonin can contrast a sun producing lots of energy by hot fusion with a sun producing only one fusion reaction a year by cold. (Koonin wanted to counter the claim that the huge pressure in the palladium lattice caused by its absorption of deuterium would somehow trigger fusion.) The choice of a salient metaphor generates humor and enables Koonin to drive home the point of his calculations to some effect.

The laughter elicited in these cases helps build solidarity and rapport. Furthermore, the humor helps to permit the speaker to make what on other occasions might be seen as inappropriate and outlandish comparisons. The metaphors may be funny, but their deployment is serious.

Using Humor to Ridicule Pons and Fleischmann's Competence as Scientists

Perhaps the most significant form of rhetoric in the cold fusion controversy has been that encountered during Koonin's and N. Lewis's talks at the APS meeting. Many commentators point to the decisiveness of these talks in initiating the demise of cold fusion. As Close remarks, "In one evening, and primarily due to Lewis's talk and Koonin's outspoken criticism of the two chemists, suddenly the emphasis had swung the other way. Now it was the skeptics who were in the leading role" (1990, 214). What is perhaps surprising about both Lewis's and Koonin's talks is that they are punctuated by humor and applause. I suggest that the talks achieve their debunking effect through the rhetorical vehicle of humor. Let us briefly examine some of the key moments, particularly from Koonin's talk (space precludes my analyzing Lewis's talk, but my initial observations are that it exhibits similar rhetorical patterns to those found in Koonin's).

In the following extract, which occurs near the start, Koonin raises the possibility that Pons and Fleischmann have committed fraud by changing their data on neutron detection between a preprint and the published version. Furthermore, Koonin suggests that the explanation for the observed data is that Pons and Fleischmann were simply observing natural radiation in their laboratory produced by radon. Koonin compares the shape of Pons and Fleischmann's gamma ray spectrum as shown in their preprint with that shown later in their published paper. The gamma ray spectrum forms the main evidence for the detection of neutrons. Any neutrons produced by the cell should be captured in the water bath by protons and thus produce gammas of a characteristic energy:

There's something funny going on with this gamma line to begin with. Here is a picture of the gamma line as it appeared in their preprint. Notice that there's a tail to the low energy side and it drops off rather steeply on the high energy side. Here is the gamma ray line as it appeared in the published article. Notice that there's a tail on the high energy side and a steep drop off on the low energy side. [Laughter. This contrast is made visible by the two graphs.] Moreover if you try to match up the points point-by-point you can't. [Distant laughter.] Even funnier about that gamma ray line is that it doesn't peak in the right place as has been pointed out by a number of people, in particular David Bailey from Toronto. I tried to indicate here as best as I could on their energy scale the location where you would expect to see the captured gamma line at 2.224 MeV. Notice that the line doesn't peak there. Rather it peaks at 2.200 or very close to that. What makes 2.200 gamma rays? Well it turns out that there is a decay of radon. Radon 222 [laughter] decays by a succession of alpha emissions to Bismuth 214, and then five percent of the time it decays to Polonium 214 accompanied by a gamma ray of 2.204 MeV.

David Bailey has estimated, as I think we'll hear from him later in the session, that the rate in a three-by-three sodium iodide should be about half of what the Pons-Fleischmann rate is reported to be. That, of course, depends on how much radon there is in their laboratory. I don't know how much radon there is in their lab, but I do know that they mine uranium in Utah. [Long laughter followed by sporadic applause, mumbled talking.]

In this extract laughter occurs at several points. The first place is after Koonin suggests that the gamma ray spectrum has been changed between the two papers. The second place is after Koonin says that "if you try to mach up the two graphs point-by-point you can't," but on this occasion the laughter is quiet and distant. The third place where laughter occurs is after Koonin announces that an isotope of radon produces the gamma ray line. Finally the punch line to the whole sequence, where both laughter and applause are generated, occurs when Koonin makes the contrast between not knowing how much radon there is in Pons and Fleischmann's lab but knowing that they mine uranium in Utah.

Although the laughter is evoked by different particular contrasts, the overall contrastive scenario is that between good science and incompetent, fraudulent science whereby the luckless chemists cannot even be trusted to rule out radiation in their own lab. The laughter generated obviously builds solidarity among the physicists. However, in this case I suggest that the more subtle aspect of humor, whereby it enables serious work to be done, is crucial. What has in fact happened is that several serious charges have been leveled against Pons and Fleischmann. They are charged with having changed data and furthermore of being so incompetent that they have overlooked an obvious source of gamma ray counts. To make such charges stick seriously one would need a lot more evidence than Koonin is able to marshall. (Later, a more powerful criticism was leveled against the neutron measurements based upon a claimed incorrect calibration of the detector.) The use of humor renders these charges credible on this occasion. The humor means that Koonin does not have to be accountable for what he has claimed; it is the veil which enables him to make serious allegations in a nonserious manner.

Koonin uses this tactic of ridicule elsewhere in his talk. In this example, Koonin is about to investigate one theoretical possibility which might enhance the fusion rate in the palladium lattice:

Let me now turn to a speculation on this first problem [of enhancement of the fusion rate in the palladium lattice]. Theorists are allowed to float trial balloons, right for those of you who are not physicists, you know, that's well accepted in the culture; experimentalists are *not* allowed to float trial balloons. [Laughter, some of its sounds exaggerated, and Koonin adds self-conscious laughter.]

In this extract we see Koonin produce a little homily about what different sorts of scientists are allowed to do. He as a theorist is allowed to "float trial balloons," but experimentalists are not. This produces much laughter. Koonin does not have to say explicitly that Pons and Fleischmann are floating trial balloons, but by implication his remarks can be heard as being about these two scientists. The humor generated helps Koonin in his task—not only by building rapport with and among members of the audience, but by allowing him to do a bit of speculative thinking himself. Again, humor seems to allow the breaching of the normal rules.

The Rhetoric of Postscripts

The two final examples of rhetoric I examine occur in the postscripts to both Koonin's and Lewis's talks, a strategy I term "breaking frame" (Goffman 1974). While the audience applauds (plus a few boos), Koonin offers an overhead transparency which stands in stark contrast to the technical transparencies he has shown earlier. It is one of Aesop's fables about the great leap at Rhodes. As soon as it is clear what Koonin is doing, it provokes laughter. Koonin reads the fable about how someone claimed to have made a great leap at the city of Rhodes and that many men witnessed it. A skeptic says "no need of witnesses; show us how far you can leap." Koonin translates Rhodes into Salt Lake City, and scribbled on the overhead is the moral—"Actions speak louder than words." This postscript in its own right generates more laughter and applause.

This rather unorthodox end to his talk works similarly to the humorous interludes throughout. The section is clearly humorous. Yet its message is again serious: Pons and Fleischmann are fools making grandiose claims which they cannot substantiate, and which any smart person can expose.

Lewis similarly tags on a postscript to his talk, running his comments about press conferences onto his formal conclusions. Again he presents his scientific conclusions that the Pons-Fleischmann cold fusion experiments are wrong.[4] He announces his postscript to laughter by saying, "If we're going to have science by press conference, these are some of the questions I would like the press to ask." A series of insinuations about probable faults in Pons and Fleischmann's experiments follows. Again, the postscript allows Lewis to make particularistic, ad hominem claims with no evidence and little argument.

Debunking Rhetoric

Perhaps the overall message of Koonin's and Lewis's talks, given their success, is that humor is a very effective form of debunking rhetoric (for more discussion of debunking rhetoric, see Pickering 1981 and Collins 1985). The laughter generated builds rapport and solidarity against its luckless victim. More subtly it allows the critic to appear to be "more scientific than thou" while engaging in the dirt and real politics of the controversy. In short, humor allows scientists to have their cake and eat it. They can accuse their opponents of breaking the scientific rules and at the same time break the rules themselves.

Conclusion

Although the rhetorical analysis of science is in its infancy and I only touch upon some of the rhetoric of this one controversy, we can see some of the subtleties of how scientists and others use rhetoric. I have addressed spoken rhetoric rather than written texts, but the work of others indicates that written texts in science and elsewhere are similarly rhetorical. Although I have not examined them in any depth in this essay, I would maintain that scientists' demonstrations also are amenable to rhetorical analysis.

By comparing aspects of the activity of science with selling my aim has not been to denigrate science. Rather my goal is to point to the rhetorical skills which underpin many human activities. If anything we should perhaps revise our view of selling and see that the skilled salesperson who demolishes the competition's product is on a par with the skilled Cal-Tech theorist who demolishes the latest experimental claims.

My claim is *not* that science is only rhetoric. In Pinch (1986), I outlined many other social processes which make up science. However, in some contexts rhetoric can be a powerful weapon. In the cold fusion case I claim that the outcome of the debate has to a large part been shaped by particular rhetorical performances.

The reason I am willing to grant such a large role for rhetoric in science stems from our revised understanding of how science works (Collins and Pinch 1993). Because no scientific experiment is alone definitive, because calibration and repeatability are in principle endlessly negotiable matters, because scientific theory is more open-ended than many think, because proofs need not be definitive (Pinch 1977), science, as an enterprise, needs rhetoric. If most scientists were not persuaded of the "truth of the matter," science would not exhibit such a degree of coherence. H. Collins (1975) suggests that a ship in a bottle is the appropriate metaphor for understanding science. Once a scientific fact (a ship) is placed in the bottle of validity it becomes difficult to see how it ever got there. One of the jobs of sociology of science is to unravel this puzzle. I suggest that rhetoric is part of the glue which holds ships together.

NOTES

I am grateful to the Cornell University Archive for the History of Cold Fusion for permission to use video materials. Video extracts are transcribed from the American Physical Society videotape of cold fusion sessions at the Baltimore APS meeting, 1 May 1989; from the American Chemical Society videotape of cold fusion sessions at the Dallas ACS meeting, 12 April 1989; and from the author's videotape collection.

1. This generalization has exceptions when taking into account variations in style between national research cultures and different fields. E.g., in computer graphics, the norm is to present a small, humorous video of the scientist's work.

2. Perhaps humor always works this way in scientific presentations and that, given the special circumstances surrounding the ACS and APS presentations, the number of such instances are exaggerated. Gilbert and Mulkay (1984) suggest that many scientific cartoons build a contrast between two repertoires of scientific discourse—the empiricist and the contingent. See also Travis (1981).

3. In the U.S. the equivalent of the Cindy doll is Barbie and her boyfriend GI Joe.

4. Interestingly, Lewis's main empirical claim, that Pons and Fleischmann had not stirred their cells and had mistaken thermal disequilibriums for heat excesses, was soon countered by Pons and Fleischmann who showed that the deuterium bubbles provided sufficient stirring. Lewis, had, it seems, based his conclusions on a series of experiments in which he had built much larger cells than Pons and Fleischmann had actually used. Like Koonin's claims over radon, Lewis's claims about lack of stirring are not definitive.

REFERENCES

American Chemical Society video of cold fusion sessions at Dallas ACS meeting. 12 April 1989. Located at Cornell University Archive for the History of Cold Fusion.

American Physical Society video of cold fusion sessions at Baltimore APS meeting. 1 May 1989. Located at Cornell University Archive for the History of Cold Fusion.

Ashmore, M., M. Mulkay; and T. Pinch. 1989. *Health and Efficiency: A Sociology of Health Economics.* Milton Keynes: Open University Press.

Atkinson, M. 1984. *Our Masters' Voices: The Language and Body Language of Politics.* London: Methuen.

Bazerman, C. 1988. *Shaping Written Knowledge: Essays in the Growth, Form, Function, and Implications of the Scientific Article.* Madison: University of Wisconsin Press.

Bloor, D. 1976. *Knowledge and Social Imagery.* London: Routledge & Kegan Paul.

Clark, C.; P. Drew; and T. Pinch. 1994. "Managing Customer Objections: A Conversation Analysis of Seller-Consumer Behavior in Real-Life Sales Negotiations." *Discourse and Society 5:* 437–62.

Clark, C., and T.J. Pinch. 1988. "Selling by Social Control." In N. Fielding, ed. *Actions and Structures.* London: Sage Publications, pp. 119–41.

———. 1992. "The Anatomy of a Deception: Fraud and Finesse in the Mock Auction Sales Con." *Qualitative Sociology 15:* 151–75.

———. 1993. "The Interactional Study of Exchange Relationships." *History of Political Economy 26.*

———. 1995. *The Hard Sell.* London: Harper Collins.

Close, F. 1990. *Too Hot to Handle.* Princeton: Princeton University Press.

Collins, H.M. 1975. "The Seven Sexes: A Study in the Sociology of a Phenomenon, or the Replication of Experiments in Physics." *Sociology 9:* 205–24.

———. 1985. *Changing Order: Replication and Induction in Scientific Practice.* London: Sage Publications.

Collins, H.M., and T.J. Pinch. 1993. *The Golen: What Everyone Should Know About Science.* Cambridge, England: Cambridge University Press.

Dear, P. 1985. "*Totius in Verba: Rhetoric and Authority in the Early Royal Society.*" *Isis 76:* 145–61.

———, ed. 1991. *The Literary Structure of Scientific Argument.*

Galilei, G. 1989. *Translation of Siderius Nuncius.* Translated by A. van Helden. Chicago: Chicago University Press.

Gieryn, T. 1992. "The Ballad of Pons and Fleischmann: Experiment and Narrative in the (Un)Making of Cold Fusion." In E. McMullin, ed., *The Social Dimensions of Science.* Notre Dame: Notre Dame University Press.

Gilbert, N.G., and M. Mulkay. 1984. *Opening Pandora's Box.* Cambridge, England: Cambridge University Press.

Goffman, E. 1974. *Frame Analysis: An Essay on the Organization of Experience.* London: Harper & Row.

Gross, A. 1990. *The Rhetoric of Science.* Cambridge, Mass.: Harvard University Press.

Hill, R.J., and K.S. Crittenden. 1968. *Proceedings of the Purdue Symposium on Ethnomethodology.* Institute for the Study of Social Change, Department of Sociology, Purdue University, Institute Monograph Series, No. 1.

Latour, B. 1987. *Science in Action: How to Follow Scientists and Engineers Through Society.* Cambridge, Mass.: Harvard University Press.

Latour, B., and S.W. Woolgar. 1979. *Laboratory Life: The Social Construction of Scientific Facts.* London: Sage Publications.

Lewenstein, B.V. 1992. "Cold Fusion and Hot History." *Osiris 7.*

Locke, D. 1992. *Science as Writing.* New Haven: Yale University Press.

Lynch, M. 1986. *Art and Artifact in Laboratory Science: A Study of Shop Work and Shop Talk in a Research Laboratory.* London: Routledge & Kegan Paul.

McCloskey, D. 1985. *The Rhetoric of Economics.* Wisconsin: Wisconsin University Press.

Merton, R.K. 1942. "Science and Technology in a Democratic Order." *Journal of Legal and Political Sociology 1.*

Mulkay, M. 1984. "The Ultimate Compliment: A Sociological Analysis of Ceremonial Discourse." *Sociology 18:* 531–49.

———. 1986. *On Humour.* Cambridge, :Polity Press.

Mulkay, M.; C. Clark; and T.J. Pinch. 1993. "Laughter and the Profit Motive: The Use of Humour in a Photographic Shop." *Humor 6:* 163–93.

Myers, G. 1990. *Writing Biology.* Madison: Wisconsin University Press.

Pickering, A. 1981. "Constraints on Controversy: The Case of the Magnetic Monopole." *Social Studies of Science 11:* 63–93.

———. 1984. *Constructing Quarks: A Sociological History of Particle Physics.* Chicago: University of Chicago Press.

Pinch, T. 1977. "What Does a Proof Do if it Does not Prove?" In P. Weingart, E. Mendelsohn, and R. Whitley, eds., *The Social Production of Scientific Knowledge.* Dordrecht: Reidel: 171–215.

———. 1986. *Confronting Nature: The Sociology of Solar Neutrino Detection.* Dordrecht: Kluwer.

Pinch, T.J., and C. Clark. 1986. "The Hard Sell: 'Patter-Merchanting' and the Strategic (Re)Production and Local Management of Economic Reasoning in the Sales Routines of Market Pitchers." *Sociology 20:* 169–91.

Shapin, S., and S. Schaffer. 1985. *Leviathan and the Air-Pump: Hobbes, Boyle, and the Experimental Life.* Princeton: Princeton University Press.

Sherry, T.F., Jr. 1988. "Market Pitching and the Ethnography of Speaking." In M. Houston, ed., *Advances in Consumer Research,* vol. 15, Provo, Utah: Association of Consumer Research, pp. 543–47.

Travis, D. 1981. "On the Construction of Creativity. The 'Memory Transfer' Phenomenon and the Importance of Being Earnest." In K. Knorr, R. Krohn, and R. Whitley, eds., *The Social Process of Scientific Investigation.* Dordrecht: Reidel, pp. 165–96.

Comment

H. Krips
Department of Communication, University of Pittsburgh

T. Pinch considers certain oral scientific presentations which employed humor not only to bring about audience laughter but also for the rhetorical purpose of shaping the audience's beliefs and desires, but he does not explain the rhetorical mechanism at work in these presentations. For example, he writes, "Laughter . . . helps build solidarity and rapport. Furthermore, the humor helps to permit the speaker to make what on other occasions might be seen as inappropriate and outlandish comparisons." This quotation leaves unclear whether the jokes (what he calls, more generally, humor) had a rhetorical effect only via the laughter they brought about:

joke → laughter → persuasion,

or whether they had a *direct* persuasive effect and also independently caused audience laughter:

joke → persuasion & joke → laughter.

He also does not explain why the audiences for these presentations laughed in the first place. As Pinch points out, such behavior is unusual in scientific meetings, "for most conference presentations by scientists . . . the audience typically applauds at the end and perhaps laugh at one standard joke."

I address these explanatory gaps in Pinch's account in the context of one of the examples he discusses: the address by Pons to 7000 chemists at the 1989 American Chemical Society conference. We must explain why the "cheap shots" which Pons directed against the physics community were (apparently) so successful rhetorically in swaying the audience in favor of cold fusion. More specifically we must explain why (1) these cheap shots were received with such unrestrained mirth by the audience; and (2) why these same bad jokes (apparently) led the audience to suspend their professional skepticism about new ideas. To simply accept as natural (as Pinch seems to) that chemists laugh publicly at other rival groups of scientists (such as physicists) and that such laughter by itself is enough to undermine their critical faculties contradicts evidence to the contrary concerning scientific skepticism about new ideas and professional courtesy displayed by scientists to other members of the academy.

I start my explanation by pointing out the complex social structure within which the cold fusion controversy was located:

A. A status hierarchy existed within institutional science which placed physicists ahead of chemists.

B. A strong professional ethos existed among chemists which contested their low status relative to physicists.

(In short, there was a contradiction between the externally and internally instituted norms governing the community of chemists.)

C. Socially instituted boundaries in science assigned certain problems to physics and others to chemistry; more specifically, the problem of producing controlled fusion reactions was assigned to high temperature plasma physics (an assignment which had serious organizational ramifications given the huge sums of money invested in fusion research by big science).

From (C) it follows that the idea of "cold fusion"—which in the context of the public fusion debate meant fusion by purely *chemical* means—not only threatened certain entrenched professional interests but also cut across traditional social categories. In Douglas's terminology, cold fusion was a boundary crosser. More specifically, the idea of cold fusion had the potential of radically undermining the extant status hierarchy (A) by implying that chemists could achieve what physicists were supposed to do but could not. The idea of cold fusion threatened to resolve the normative contradiction between (A) and (B) in favor of (B) and the professional interests of chemists.

All of this effected a transformation of cold fusion into what was, for chemists, an object of desire and therefore also an object of belief. Correspondingly, of course, it was transformed into an object of loathing and disbelief for physicists. I here rely upon the work of Lacan (1981, chaps. 5, 9) on the constitution of desire. Lacan, following Freud, argues that new objects of desire form in response to and as screens for "gaps" in the lives of human subjects. Such gaps are always repetitious (*Wiederholungenen*) of a repressed primal structure of loss, which may manifest in various ways—for instance, as a contradiction in the normative framework within which subjects operate.

The chemists' laughter, I suggest, expressed pleasure produced by the experience of such a new desire and the promise of its gratification. This expression of pleasure took the form of laughter because of the special form of Pons's address. It tells the truth that chemists run a poor second to the well-endowed physicists (look at the small size of Pons and Fleischmann's apparatus for bringing about cold fusion compared to the giant coils of the physicists' fusion machines). But it is also clear that truth telling truth here is intended deceitfully, since the implication is that chemists' lack is really their strength. (Pinch describes the joke form as "Two different frames or perspectives . . . brought into unusual conjunction.") Thus Pons's address belongs to a special category of jokes singled out by Freud, namely, those which show a speaker (who may or may not be the joketeller) deceiving an audience by the paradoxical means of telling them the truth.[1]

I suggest that the communal nature of the laughter created by Pons's address not only expressed a desire for cold fusion (i.e., fusion for chemists) but also functioned as part of the mechanism creating the desire. I have in mind Durkheim's idea that communal "effervescence" not only expresses but also creates group solidarity by establishing a common bond to certain communal objects of desire.

In sum, we can explain the rhetorical effects of Pons's public evocation of cold fusion as the result of a dual mechanism: First, the jokes he employed constructed cold fusion as an object of desire by presenting it as a resolution of a contradiction lived by the audience; second, his address brought chemists en masse into a situation where they communally entertained and responded to an idea of something which they desired, thereby reinforcing each other's desire and certain corresponding beliefs. At the same time, we can explain the general mirth which greeted Pons's cheap shots. It was not so much the result of a low sense of humor among chemists as an expression of the pleasures associated with a new object of desire.

NOTE

1. "Is it the truth if we describe things as they are without troubling to consider how our hearer will understand what we say . . . . I think that jokes of that kind are sufficiently different from the rest to be given a special position" (Freud 1950, 115). Another example of such a joke, discussed by Lacan, is the famous "Yes, why do you lie to me saying you're going to Cracow. . . . I should believe you're going to Lemberg when in reality you *are* going to Cracow?" (Lacan, 1972, 49).

REFERENCES

Freud, S. 1950. *Jokes and Their Relation to the Unconscious*. New York: Norton.
Lacan, J. 1972. "Seminar on the Purloined Letter." *Yale French Studies* 48: 38–72.
———. 1981. *The Four Fundamental Concepts of Psychoanalysis*. New York: Norton.

8

American Intransigence

The Rejection of Continental Drift in the Great Debates of the 1920s

Robert P. Newman

Department of Communication, University of Pittsburgh

The Orthodoxy

T. C. Chamberlin was arguably the most prestigious figure in American geology during the first two decades of the twentieth century. Chamberlin's reputation was well deserved. His work on the Wisconsin Geological Survey led to great advances in glaciology and the first identification of fossil coral reefs in the United States. When invited, he headed the geology department of W. R. Harper's new University of Chicago, teaching there until the end of his career, founding and editing the *Journal of Geology*.

Chamberlin's most enduring single publication was probably his *Science* article, "The Method of Multiple Working Hypotheses," published in 1890, reprinted in many journals, and available as late as 1977 from the American Association for the Advancement of Science (AAAS) as a reprint (Chamberlin 1890). This article strongly attacks the tyranny of dominant hypotheses. Even today, readers of this article finish with a determination to rethink their research to see what damage theory-induced bias has inflicted.

One of the tragedies of American science is that the great man who produced such sage advice, who applied it to analyses of North American glaciation and to the regnant nebular hypothesis of the origin of the earth, became so enamored of his own planetesimal hypothesis with its doctrine of the permanence of continents and oceans that he closed his mind to any alternative and led his son and professional colleagues to a dead-end fanaticism. As R. Dott (1992) says, Chamberlin's papers "reveal a great love affair with his planetesimal theory characterized by elaborate, rhetorical pirouetting with old and new evidence" (p. 36).

When Chamberlin spun out the implications of the planetesimal hypothesis for geology, in a series of *Journal of Geology* articles under the title "Diastrophism and the Formative Processes" and in Carnegie Institution of Washington annual reports, he concluded that the earth was now shrinking under gravitational forces, which was a major cause of mountains.[1] The earth had a solid interior, without the molten layer postulated by some geophysicists, but radioactive elements created pockets of magma. This was true because the planetesimals forming the earth had been drawn from the sun by a passing star, the bolt of sun materials was not uniform in composition, and the unevenness was retained in the planetesimals and in the earth's structure. Continents were composed of lighter materials than ocean basins; this accounted for their greater height. Isostatic adjustment of the solid but plastic interior allowed this, just as gravity allowed the solid ice of glaciers to move. Sinking of the heavier ocean basins exerted horizontal pressure on the continents, squeezing them upward and contributing to mountain building. By 1920, this scenario of earth formation and mountain building was largely complete; it provided no latitude for mountain building by continental movement.

Chamberlin's dogmatic rejection of any form of mobilism seems inconsistent with many of his other beliefs. As Brush (1978, 14) notes, he flirted with mobilism in an 1896 lecture, and his papers at the University of Chicago Regenstein Library contain a reference to "intercontinental migration tracts" dated 12 March 1923 (Chamberlin Papers, Box 8, addenda folder L). But when mobilist hypotheses became an issue in American geology, Chamberlin hardened.

Of Chamberlin's followers, the one of greatest moment in the drift debate was B. Willis of Stanford. Willis, born fifteen years later than Chamberlin, sought the advice and support of the Chicago geologist from the beginning of his professional career, and continued contacts with Chamberlin's son Rollin after the older Chamberlin's death. Central to the Willis-Chamberlin bond was acceptance of the planetesimal hypothesis, with its denial of a uniform subcrustal layer of molten magma, its hypothesis of mountain building by pressure from the oceans, and hence its commitment to the permanence of continents and oceans.

The final American geological mandarin who played a major role in the drift debate was C. Schuchert of Yale. A noted paleogeographer, Schuchert did not accept the planetesimal hypothesis, holding that the earth at one time was molten. Chamberlin's writings on the criteria for establishing geologic time periods, however, were of great use to Schuchert, whose commitment to permanent continents and oceans was also total.

While Chamberlin, his son, Willis, and Schuchert were the prime activists among American geological elites in the crusades against mobilism, they had support in depth. In the crucial 1922 to 1933 period, six of the twelve presidents of the Geological Society of America (GSA) were on public record in their presidential addresses as opposing drift; of the other six, only one, R. Daly of Harvard, took a mobilist position. Of geologists elected to the National Academy of Sciences (NAS) during this period, ten were active opponents of mobilism, *all of them committed before the notorious American Association of Petroleum Geologists (AAPG) debate.* The only exception among NAS members, again, was Daly. Not all the opponents of drift agreed as to how the continents got where they are, whether the mountains were raised by shrinking of the earth's crust, or how life forms got distributed as they were. But all the vocal opponents of mobilism were elites, and they stood united (except, again, for Daly) against heresy.

Permanentism was clearly orthodoxy in early twentieth-century American geology. J. D. Dana's version of the contraction hypothesis explained well enough what structural geologists thought they saw in the Appalachians and Rockies. Schuchert's land bridges, dispersal by rafting, and other strategems satisfied paleontologists. Subsiding ocean basins pressing against restraining continents explained orogeny. And Chamberlin had, as R. M. Wood puts it, "implicated the nation in his theories, [so that] to disagree with him was to countenance that worst of all American sins, antipatriotism . . ." (1985, 32).

The Heresy

Mobilism was not entirely unknown in 1920 America. F. B. Taylor (1910) offered the first embryonic exposition of a mobilist position.

He presented a picture of continental movement away from the poles, accounting in this fashion for the east-west trending of the Alps and Himalayas. His later writings attributed this movement to capture of the moon by earth in the Cretaceous period, causing tidal forces sufficient to crumple continents that had previously been closer to the poles. Taylor made no attempt to consider the full range of geophysical, paleontological, paleoclimatic, and geodetic evidence later mustered by Wegener. As Hallam notes, "In this respect the relationship of Taylor to Wegener is not unlike that of Wallace to Darwin" (1973, 6). Despite Taylor's standing in the "upper-middle echelons of the geological community" (Laudan 1985, 118), Taylor's mobilist theories were ignored by most American geologists. (On Taylor's mobilist views, see Laudan 1985; Wood 1985, 48–50; and Marvin 1973, 63–64).

Then there was A. Wegener. His development of the displacement hypothesis (mistranslated as continental drift, which label stuck) is well known). Wegener's first formulation was in 1912, and largely failed to migrate beyond continental Europe (see Jacoby 1981). The first world war prevented knowledge of the first and second editions of *Die Entstehung der Kontinente und Ozeane* from reaching the English-speaking world. His third edition, in 1922, was published in English translation (London and New York) in 1924; but the doctrine preceeded the translation (Wegener 1924).

The first mention of Wegener's displacement hypothesis is an American periodical was by O. Holtedahl, a Norwegian writing in the January 1920 *American Journal of Science* (*AJS*). Holtedahl (1920) discussed the structural similarities on the opposing sides of the North Atlantic and the forces that might have created the structure. He had read Wegener, and much of his article is in tune with Wegenerian mobilism.

Holtedahl noted that most American geologists believe in the permanency of the "great features of earth relief" (ibid., 21) and he cited Willis and Schuchert, among others. But he questioned permanentist doctrine, "It certainly will be hopeless for the advocates of permanency of the oceanic basins to apply the theory to the Norwegian Sea" (ibid.). I have found no response to Holtedahl's article, even though *AJS* was a widely read periodical. The name Wegener, at that time, stirred no spirits from the vasty deep. The next reference to drift and

Wegener came a year later, again from a European, again as an after-thought (Brouwer 1921).

In 1922, Wegenerian ideas significantly penetrated American geology. The first discussion was a brief note in the *Pan-American Geologist* (*PAG*), a proprietary publication in Des Moines, Iowa run by editor C. Keyes. Keyes had great difficulty during the decade deciding what to make of mobilism: In the February 1922 issue he carried an article by C. K. Leith (a future president of GSA). This early in his career, Leith was heretical, "Under its environmental forces the solid earth, the popular symbol of strength and permanency, proves weak and incompetent. . . . Much of the sharper diastrophism seems to be confined to a thin surficial zone. Deeper movements of a more massive type, periodic and possibly slower, seem to be implied by the relative movement of the great earth segments as represented by the continents and ocean basins" (1922, 85). Wegener was not mentioned.

Also in February 1922 subscribers to *Nature* saw a brief anonymous review of the second edition of Wegener's *Entstehung*. It was friendly, but sounded alarms, "This book makes an immediate appeal to physicists, but is meeting with strong opposition from a good many geologists. This opposition is to be expected, for the author replaces the whole theory of sunken continents, land bridges, and great changes of earth temperature by a displacement theory. . . . The revolution in thought, if the theory is substantiated, may be expected to resemble the change in astronomical ideas at the time of Copernicus" ("Wegener's Displacement Theory" 1922).

Schuchert heard the alarm. His lead article in April's *PAG* was a 21-page translation of C. Diener's 1915 attack on Wegener (Diener [1915] 1922). Diener was a prominent Viennese paleontologist, and surprisingly for an Alpine scholar, he was highly conversant with American doctrines. Diener rejected Suess' mobilism; he emphasized the permanence of continents and oceans, going back to the American Dana for the lineage of permanentist theory and frequently citing later American writers.

Diener's article fit Schuchert's beliefs like a glove; Schuchert's (1928) later criticism of Wegener in the AAPG publication on continental drift bears strong resemblance to Diener's earlier work, with one major difference: By 1928, Schuchert could no longer blitherly assert, as Diener did in 1915, that "[no] geophysical argument is op-

posed to the sinking of individual fragments of the continental plat-
forms into the abyssal deeps" (p. 197). The Diener translation was
Schuchert's first full-scale attempt to warn his fellow geologists against
Wegener's pernicious doctrine.

While American geological elites began to grapple with mobilism,
many Alpine geologists found that only a mobilist hypothesis could
explain the vast overthrusts they confronted daily. Foremost among
them was the Swiss tectonician E. Argand. Defying a wartime (1914–
1918) Swiss ban on possessing German literature, Argand had managed to obtain Wegener's early writings, and in 1916, at the Neuchatel
Society of the Natural Sciences, he "presented Wegener's ideas as a
surprise" (Carozzi 1977, xvii). He then began to revise his tectonic
map of Eurasia along mobilist lines. Fanatical social constructionists
will say that what Argand now saw in the Alps and Himalayas was
"constituted" by his new hypothesis; geologists who knew Argand
were convinced that, having found permanentism an inadequate approach to his data, he simply adopted the new hypothesis as superior.

Argand gave his first full mobilist presentation of the world's great
mountain chains as the inaugural address at the Thirteenth International Congress of the Geological Sciences in Brussels on 10 August
1922. It is not Whig history to say, as J. F. Dewey does, that Argand
produced one of "the two works in tectonics that have most changed
our understanding of the earth's tectonic behavior . . ." (1978, 191).
The American geologists' disregard of Argand in the 1920s is nothing
less than a scandal. Of twenty-five Americans registered for the Brussels Congress, five were official GSA delegates. Of those present, only
Daly clearly appreciated and acknowledged the Swiss genius.

The only references to mobilism in American literature during the
closing months of 1922 were with one exception bitterly hostile. GSA,
holding its annual meeting in Ann Arbor 28–30 December 1922, appropriately scheduled a Friday afternoon to hear papers and discussion on dynamical, structural, glacial, and physiographic subjects.
Daly of Harvard was the only "heretic" for whom space was provided.
Daly (1923) used temperature gradients in the crust and seismograms
to conclude that under the crust lay a noncrystalline, elastico-viscous
substratum. The secretary for this meeting noted that Daly offered an
hypothesis of the continents and ocean basins based on the Taylor-
Wegener hypotheses of the horizontal movement of continents. Arti-

facts of this meeting are slim, but the subsequent *GSA Bulletin* reveals that Daly's remarks were greeted with sarcasm and disdain.

At the same GSA convention, A. Keith, five years later to serve as GSA president, delivered before the Society the first sustained and comprehensive attack on mobilism. Keith emphasized permanence, the absence of hard data supporting continental displacement, the implausibility of Wegener's displacing force, the superiority of crustal shrinking to account for mountains, and a new factor: "[Wegener's] hypothesis apparently must, therefore, fall back on a cause which is of the order of the special convulsion of nature appealed to in the early stages of geologic work. Such a convulsion, which affected a whole hemisphere, could have no probable cause except one of an astronomic nature" (1923, 361). Wegener had not even hinted at a "convulsion of nature" to account for displacement, much less an astronomic event. Keith sensed that mobilism might be vulnerable to attack on catastrophic grounds despite what Wegener wrote, and perhaps he was right. Several observers of this period believe that part of the reason for mobilism's rejection in the United States was having been tarred with the brush of catastrophism. The year thus ended with mobilism known and feared; the crusade against it was already begun.

In 1923, T. Chamberlin directed a full measure of scorn at old and new forms of mobilism. Invoking Shakespearean personae (the two Dromios in *Comedy of Errors*) Chamberlin, in a *Journal of Geology* editorial, imagined a conversation between two geologists. One offers to enlighten his mate by explaining the origin of the ocean basins. The other says "fire away," whereupon the first presents a bowdlerized version of drift theory. This story, said Chamberlin, was told and retold for years, until it "lost flavor"; the story was then dropped. The denouement: "We learn, however, that it has broken out again in a new and virulent form in the land that is alleged to have been most queered by the original catastrophe" (1923, 153). In his earlier contention with Lord Kelvin, Chamberlin had directed his ire straight at his target. Wegener was beneath acknowledgement.

The contempt of the mightiest of American geologists notwithstanding, a significant discussion of mobilism took place a month later at a joint meeting of the Washington Academy of Sciences and the Geological Society of Washington, 18 April 1923. The program consisted of three addresses on the "Taylor-Wegener Hypothesis": 175

people attended ("Proceedings: Washington Academy of Sciences" 1923).

The opening speaker, gaining what was probably his best opportunity to promote his own variety of mobilism, was Taylor. We do not know how the audience responded.

Daly, undeterred by his recent pummeling at Ann Arbor, followed Taylor. After outlining Wegener's scenario, he notes its difficulties. Wegener's "assumption of practically no strength in the sub-oceanic crust" (ibid., 477) is indefensible. Daly offers other criticisms, then comes to his bottom line, "Wegener's presentation of the case has provoked much discussion and many objections, in addition to those listed. So obvious are his logical inconsequences and his failure properly to weigh ascertained facts that there is danger of a too speedy rejection of the main idea involved" (ibid., 448). Daly then suggested his own version of "continental sliding", which involved a concatenation of factors too complicated to present here.

Whether the auditors in Washington were less critical or more polite than those in Ann Arbor is unknown. However, we know that in this, the second major American symposium on mobilism, the permanentists were not in control.

American periodicals conducted a crusade against mobilism in 1924. Seven considerations of the subject were published; not one was favorable. *AJS* led the list of attackers, with four articles attacking drift. *Journal of Geology, Science,* and *PAG* each had one.

Two major happenings in 1925 brought the simmering mobilist controversy to a boil: The English translation of Wegener's third edition reached the United States and was reviewed; and W. Bowie (1925c), with his official status on the Coast and Geodetic Survey, authored a spectacular article in the *New York Times* reporting that the world's geodesists were starting a longitude-measuring project to test Wegener's hypothesis. Despite Dana, despite Chamberlin, despite Willis, Schuchert, Berry, and all the GSA presidents who opposed drift, the three reviews of Wegener's book were tolerant.

One should not assume from these reviews that the enemies of Wegener had disappeared. The absence of *any* reference to the English translation in the journal of the geological establishment (*GSA Bulletin*) and the Chamberlin proprietary (*Journal of Geology*) speaks eloquently.

More eloquent was Bowie, Chief of the Division of Geodesy, United States Coast and Geodetic Survey. In 1925, Bowie put out a series of reports on the Madrid meeting of the International Geodetic and Geophysical Union, held the previous September and October. This meeting completed plans for a world-wide longitude net, not just *scientia gratia scientiae,* but explicitly for testing Wegener's continental displacement hypothesis. Wegener assumed continuing movement of the Americas and Greenland away from Europe. Because previous measurements of Greenland's longitude had been calculated on the basis of imprecise moon observations, scientists would now use highly accurate radio telegraphy.

Peculiarly, Bowie, who a year later would join the Wegener-bashers at the AAPG symposium, gives no hint in his Madrid reports that Wegener was not in good standing with American geologists. Indeed, he emphasizes that only acceptance of the geodesists' favorite hypothesis, isostasy, made Wegenerism possible. The first of Bowie's reports was in the *AJS* February issue (1925b); the second in the June issue of *PAG* (1925a), and the third, three times as long as the first two, *occupied almost a full page of the New York Times* on Sunday, 6 September 1925 (1925c). The scarehead read "Scientists to Test 'Drift' of Continents." The subhead was "Observations of Longitude and Time at Many Points Will Determine After a Period of Years Whether Land Masses Move Like Icebergs in the Sea—Radio to Carry Signals—An International Experiment."

The *Times* editorial page on the same day, under the heading "Drifting Apart," further legitimizes mobilism by noting that at a scientific meeting in London, J. W. Gregory of Glasgow endorsed the geodesists' test. Furthermore, the new instrument, radio, will solve the geological puzzle and "help to hold the nations together spiritually as their lands were once held physically in the giant continent . . ." ("Drifting Apart" 1925).

No other blockbuster comparable to the Bowie message surfaced during 1925, but some items are noteworthy. *AJS* and the *AAPG Bulletin* carried stories about mobilism, the latter definitely favorable (Thornburg 1925). More comforting to the permanentist camp was the Smithsonian *Annual Report for 1924,* published in 1925, carrying P. Termier's (1925) oft-quoted lecture to the Institut Oceanographique of Paris. This is a piece of French artistry to match the elan of Argand's masterpiece. Termier, however, head of the French Geological Survey, was no convert to Wegenerism. His seventeen pages cover all the issues, chiding and praising Wegener, emphasizing the importance of tangential movements in mountain building, but saying that Wegener claims too much for them, and admitting his temptation "to respond in the affirmative" (ibid., 224) to Wegener's rhetorical questions, then backing off when he feels that the mobilist agenda is too convenient.

Termier was later quoted illegitimately by Schuchert in the AAPG symposium. Schuchert (1928) takes from Termier these skeptical lines about Wegener's hypothesis, "It is a beautiful dream, the dream of a great poet. One tries to embrace it, and finds that he has in his arms but a little vapor or smoke; it is at the same time both alluring and intangible" (Termier 1925, 236). Termier was too wise to quit here. Only Schuchert's bias kept him from acknowledging what Termier said next:

But in all reality we can not conclude, we can not say, that there is really nothing in Wegener's theory; neither can we affirm that it does not contain some truth. Our knowledge is very limited. It is always necessary to close a lecture on geology in humility. On the ship earth which bears us into immensity toward an end which God alone knows, we are steerage passengers. (Ibid.)

The American reactionaries who refused any cogency to mobilism never closed their lectures in humility.

The AAPG Symposium

On 15 November 1926, so the histories tell us, the fate of Wegener's ideas in the United States was settled at the New York AAPG symposium. LeGrand is just one, and not at all the most misleading, of the writers who exaggerate the impact of this meeting, "The arguments mounted and the positions taken in that debate helped set the pattern for the North American response for the next four decades" (1988, 64). As should be clear by now, the arguments against Wegener had been articulated publicly before then; dozens of articles had appeared pro and con; perhaps not all Americans had made up their minds, but none of the opinion leaders who attended the AAPG symposium came to learn. W. van der Gracht, the only mobilist who took part, did not play on a level field.

The symposium was entirely due to van der Gracht's vision and energy. Van der Gracht was born in Amsterdam in 1873, received a Doctor of Law degree from Amsterdam University in 1899, and a degree in mining engineering from Freiberg in 1903. In 1905, when only thirty-two years old, he became Director of the Geological Service of the Netherlands, holding that post until 1917. During this period, he consulted for mining projects worldwide, including a year as advisor to the Netherlands Colonial Expedition to the Dutch East Indies. In 1917 van der Gracht came to the United States as president of the Roxana Petroleum Company, a Royal Dutch Shell subsidiary in Oklahoma; in 1922 he moved to the Marland Oil Company of Texas.

During his first year in the United States, van der Gracht participated in founding AAPG, which numbered among its members several of the geological elite: Schuchert, J. T. Singewald, D. White, and A. Lawson. Willis was not a member, but Willis's two sons worked in the Texas oil fields, one of them under van der Gracht's supervision; hence Willis and van der Gracht became more than casual acquaintances. Probably at van der Gracht's suggestion, Willis was invited to speak on crustal deformation at the Denver meeting of the AAPG in September 1926.

Van der Gracht and Willis exchanged several letters in the summer of 1926. The earliest to survive is dated 23 August; van der Gracht wrote Willis from a field trip in Canada, responding to Willis's request for advice about geological contracts in Java, which Willis was about to visit on a round-the-world trip. In four handwritten pages, van der Gracht described the wonders of the Dutch East Indies and advised how Willis could locate geologists there (Willis Papers (WP), Van der Gracht to Willis, 23 August 1926).

Six days later Willis wrote van der Gracht a long reply. After three paragraphs of pleasantries and thanks, Willis got around to anticipating the East Indies. His response to van der Gracht's mobilism hardly squares with his later conduct:

In approaching these new scenes I try to hold myself free to accept the facts with an unbiased mind. I grant you that I find it somewhat difficult for a man of my age has usually thought himself into intimate relations with some preferred hypothesis and I am no exception to the rule. I know what I expect to find. If, however, the facts require a different interpretation I shall try to be ready to follow them wherever they may lead.

.... I am prepared to listen to argument and to weigh evidence in discussing it. (WP, Willis to van der Gracht, 29 August 1926)

The letter shows Willis in a cordial dialogue with van der Gracht, entirely free of the fanaticism he showed years later.

Willis's paper at the Denver AAPG meeting discussed the two types of diastrophism he saw in North America: folding and shearing. Both cause significant crustal movement, and the first part of Willis's paper seems to build up to a mobilist conclusion. After all, the Appalachians represent at least a 35-mile shortening (and hence movement) of the crust, and the Santa Ynez range in California moved 24 feet in a mere 30 years. But no,

[in] thus recognizing very large horizontal movements, I lay myself open, considering the modern trend of thought, to the suspicion that I accept the views of Wegener. But I beg to point out that the displacement, as I have described it, requires a movement from the ocean against the continent, and, if we take account of both shores of the Atlantic, implies an expansion of the sub-oceanic mass. This is not the concept of continental drift. (Willis 1927, 38)

When van der Gracht commented on this paper, he complimented the lucid picture of American folding and shearing, and modestly expressed his opinion:

When, however, we come down to primary causes, we differ. Although Bailey Willis accepts great horizontal movements, particularly in the deep basement, as I do, he is very careful to make it clear that he does not want to be suspected of adhering to Wegener's theory of relative continental drift. . . . (Ibid., 46)

Van der Gracht did his best to assure comprehensive discussion of continental drift at the New York Symposium two months later. The *AAPG Bulletin* in October carried an extensive notice inviting participants to speak. The actual discussion on Monday night, 15 November 1926, in the auditorium of the Engineering Societies Building, took place with van der Gracht giving an opening presentation and five of the most prestigious American geologists attacking him in often impolite terms—via Wegener, of course.

Every one of the Americans who actually spoke was a former or future president of GSA. Every one of them had been, or was to be, elected to NAS. Every one of them was anti-Wegener. Passionate speakers are often said to talk to the already converted, wasting their time. Van der Gracht's burden was that he talked to the never-to-be converted.

Because the original invitation drew only antimobilist speakers, van der Gracht probably extended a general invitation to submit papers for publication at the close of the New York meeting, hoping that at least a few mobilists would respond. Only Taylor obliged. The other five Americans who responded reinforced the five who had spoken. Van der Gracht knew, and said, that Americans were unfriendly to mobilism, but he must have grossly underestimated their hostility. One wonders why van der Gracht did not approach Daly, or if he did, why Daly failed to respond.

Van der Gracht likely solicited personally the papers from Europeans, especially Wegener. The Wegener contribution was unfortunate: brief, avoiding the issue of force that was uppermost in American minds, and bereft of the mobilist support he could have mustered from British, Swiss, Dutch, and South African geologists of great eminence. Wegener did cite Bowie's article in the *New York Times,* and lauded the international project to measure longitudes, but if he thought that data yet far in the future would cool the mounting hostility to drift in the 1920s, he was wrong. The only other point he touched on was the argument against his claimed distribution of ancient tillites; here he won points for courage and perspicacity, but not for persuasiveness. The great Canadian glaciologist, Coleman, as well as the Chamberlins, simply said "No."

The AAPG symposium did not attract attention at the time. We do not know how many attended. This is somewhat surprising, as the *New York Times* had a month earlier carried another prominent article on the effort to measure continental movement (Boone 1926). This article quoted Bowie, who again asserted, "The earth's crust is resting on material which seems as if it were plastic. The crust floats on this material as the great ice field of the arctic floats on the waters of that ocean" (ibid., 4). One would have expected Bowie to tip the *Times* reporter off to the upcoming great debate. Instead, the *Times* covered a meeting of the International Conference on Bituminous Coal in Pittsburgh held the same day as the AAPG symposium.

The AAPG symposium was not the only activity on the mobilist front in 1926. *Nature* continued coverage of the ongoing debate in Britain. The British *Geographical Journal* for April presented an extensive account of L. Collet's (1926) lecture, with audience questions and comments afterwards. That affair was civil and heuristic in contrast to similar American discussions. *PAG* carried S. Richarz's (1926) negative review of the Koppen-Wegener book on paleoclimates, but followed this in April with H. Brouwer's (1926) strongly pro-Wegener article. *Scientific American* carried A. Ingalls's (1926) article which took no sides. Mobilism had not yet been "killed."

The Willis-Schuchert Crusade

Willis was at the height of his influence in 1927. Just back from his world tour of volcanic-earthquake zones, scheduled to assume the presidency of the GSA in December, holding a commitment from *Scientific American* for an early article on his volcano findings, he was determined not to let the mobilists gain influence in American geology. He was particularly disturbed at publication by the Carnegie Institution of Washington in 1927 of A. du Toit's article comparing the geology of South America with that of South Africa (du Toit and Reid 1927).

The first salvo in Willis's (1928c) attack on the arch heresy came in the February 1928 *Scientific American*. It was directed specifically at Wegener, J. Joly, and Daly, though none were mentioned by name. The sarcasm he so liberally directed at mobilists was much in evidence:

Our Mobile Earth is the title of a widely read book by a distinguished scientist [Daly] who leans to the extreme idea of a plastic globe on which the continents have drifted apart, floating in a substratum that is conceived to resemble sealing wax in its capacity to yield gradually to small but continuing forces. Others, of who I am one, incline to the theory of a good, solid Earth . . . a certain type of mind soars into the realms of speculation. . . . A few facts suffice to equip his airplane; he rises on the wings of a plausible argument; but he usually falls far short of the discovery of Truth. A recent work which would actuate the career of the Earth by the radio-activity of rocks, regarding which our knowledge is inadequate, belongs in this class. We may admire the intrepidity of these intellectual aviators, but we fly with them only at the risk of a crash. (Ibid., 151)

In March 1928, one month after Willis's *Scientific American* article appeared, the AAPG publication *Theory of Continental Drift* (TCD), which purported to be the papers given at the New York symposium, came out in Tulsa. Much has been written about this document and no extensive account of its substance will be attempted here. The tone, the rhetorical style, and the invective are of immediate concern.

TCD begins with van der Gracht's lengthy and moderate argument for some form of mobilism. None of the adjectives applied to Wegener and du Toit (dogmatic, doctrinaire, fanatical) can possibly be charged to van der Gracht. Possibly because of his reasonableness, he gets short shrift from most commentators on the symposium; they concentrate on the colorful charges of the permanentists.

Willis is first among the permanentists, with a relatively brief (six-page) contribution. His first four pages are as moderate as van der Gracht's. He points to difficulties in Wegener's mechanics, offers "land bridge" explanations of faunal migrations between continents, and claims that such affinities are insignificant. Toward the end of his article, however, he gives his disparagement of Wegener. The "manner in which the theory is presented" (1928a, 82) is all wrong. He claims Wegener presents no direct proofs, which is a misrepresentation; the Greenland measurements were direct, though not conclusive. Willis accuses Wegener of circular argument, but Wegener's argument is not circular. He accuses Wegener of ignoring facts and principles opposed to drift when Wegener goes out of his way to cite and acknowledge his opponents; Willis is the one who ignores the findings of the prominent European mobilists. Finally, Wegener's book "leaves the impression that it has been written by an advocate rather than by an impartial investigator" (ibid.). No controversialist can avoid advocacy, least of all Willis. He wrote all of his works on earth dynamics as an advocate.

R. Chamberlin was next. Unlike his father's *Journal of Geology* editorial, R. Chamberlin's (1928) sarcasm is out in the open. He begins with a rhetorical question, "Can we call geology a science when there exists such difference of opinion on fundamental matters as to make it possible for such a theory as this to run wild?" (ibid., 83). No argument there, just name calling.

We come then to Schuchert's (1928) 41-page essay. Much of it comes from his translation of Diener for *PAG* in 1922, but he now goes beyond Diener. His selectivity of evidence will be discussed; here let us note the unscientific rhetorical strategems. More than other symposiasts, Schuchert invokes the false dilemma. If drift tore the Americas from Eurafrica, as Wegener claims, the coastline-continental shelves would be splintered, eroded, and distorted. Hence Wegener's claim that South American coasts match African coasts proves *that drift has not occurred* (ibid., 1928, 113).

Demeaning phrases occur throughout the Schuchert piece: Wegener's maps are not worth the paper on which they are printed; statements of support by J. W. Evans in his introduction to the English edition of *Entstehung* are astonishing; Wegener's use of the torn-newspaper analogy is absurd, but also characteristic; the "nimbleness and versatility" (ibid., 134) of Wegener's mind enables him to "make all facts fit his hypothesis" (ibid.). Worse, Wegener "generalizes from the generalizations of others" (ibid.). So do we all, including Schuchert, and two pages later he enlightens us with another bit of epistemological lore directed at Wegener, "Facts are facts, and it is from facts that we make our generalizations, from the little to the great, and it is wrong for a stranger to the facts he handles to generalize from them to other generalizations" (ibid., 139). It would be overkill to show the extent to which Schuchert is hoist by his own petard. Schuchert cannot be excused for doctoring the Termier quotation cited earlier.

The AAPG symposium came from the press with a 75-page introduction and a 30-page conclusion by van der Gracht, other prodrift statements by Taylor and the four Europeans, skeptical but rational antimobilist statements from four Americans, and strong anti-Wegener statements by Willis, R. Chamberlin, Schuchert, and Berry. The space allocation is hardly against Wegener:

Prodrift pages	141
Moderate antidrift	29
Strongly antidrift	56

Thus the symposium as printed carried a clear preponderance of prodrift material, 141 pages to 85 negative. Why then, have so many analysts credited the symposium with such an impact on American thought? And if the symposium was that influential, whose contributions were so persuasive? No answers appear to these questions. All we know is that the first printing sold out quickly.

Publication of the AAPG booklet did not still discussion of mobilism. Daly brought Collet, the outspoken proponent of mobilism at Geneva, to Harvard for the winter terms of 1928 and 1929. A. Holmes (1928) wrote a favorable three-page review of the AAPG symposium in *Nature*. And Willis clearly did not believe that the AAPG publication was the final word. He began to plan further efforts against mobilism for the GSA meeting in December.

First and foremost at the GSA meeting was his presidential address (Willis 1928b). The speech is vintage Willis. He credits eight geologists with having furnished "working concepts"; all are Americans. His historical sketch of ideas of continental genesis begins with Buffon, who, according to Willis, saw "strange things, even in the morning light" (ibid., 286). Some other writers did not impress Willis, "Suess, Haug, and others of the modern school . . . also see strange things, even in the morning light" (ibid.). He admits to some radioactivity in the earth, concentrated in the outer shell; but there can be no accumulation of heat beneath the continents (ibid., 298). R. Chamberlin is credited with showing that the difference in the two sides of the earth, the water hemisphere and the land hemisphere, are "inherited from the difference in density of the two ends of the bolt that was shot from the sun to form the planet" (1928b, 319). These are the opinions of a geologist who considered mobilist ideas as poppycock.

The major American journals generally ignored mobilism in the months following Willis's effort, with one exception: *Journal of Geology* in May 1930 carried a three-page review of the AAPG symposium by F. A. Melton of the University of Oklahoma. The first paragraph of this review puts AAPG in an unusual light:

A prominent professor in a well-known eastern university, while talking to the reviewer recently, commented on the sense of shame in their profession which must have moved certain members of the American Association of Petroleum Geologists and certain oil companies to secure the publication of this book. To him it seemed a weak attempt to justify the existence of the Association. The reviewer, on the contrary, believes that its publication was a praiseworthy undertaking and a good use for surplus funds. He hopes the Association will sponsor other publications from time to time. If a sense of shame is to be identified in any way with this published collection of papers, it will probably be experienced at some time in the future by a few of the well-known contributors who have speculated so freely about the conditions within the surface of the earth. (Melton 1930, 284)

Melton does not tell us who he thinks will look ashamed because of their speculations. Perhaps some geological elites may have looked down their noses at the AAPG as a mere trade association, but nothing was second class about the symposium participants. If the symposium was as deadly for mobilism as most analysts maintain, it was because the most virulent attacks on Wegener and drift came from opinion leaders.

Melton evenhandly presents pros and cons of drift; his only strong objection is the extent to which some of the participants asserted isostasy to apply even to areas of very small size. Bowie and White are the culprits. Indeed, Bowie's devotion to a concept of isostasy based on the Mohorovicic Discontinuity was an unfortunate development. As G. Friedman and J. Sanders put it:

The dominance of the geophysical theories based on the model of a uniform mantle lying below a crust of varying thickness, Sialic beneath continents and Simatic beneath ocean floors, with all adjustments taking place just beneath the M-discontinuity, made it impossible to propose any acceptable mechanism that could explain continental drift. Alfred Wegener . . . argued his cause in vain because he could invent no plausible mechanism to explain the motions that he was so sure had taken place. (1978, 437)

Willis's public comments about mobilism were acerbic, but in his letters to du Toit (as to the mobilists van der Gracht, Holmes, Daly, and Molengraaf) he wrote in a different tone. However open-minded his letters to mobilists, his major preoccupation for the next two years demonstrates his undying devotion to the cause of permanentism.

Willis chose Schuchert for a comrade-in-arms. In some two dozen letters between them during 1931 and 1932, they hammered out their doctrinal differences and planned a campaign to satisfy both paleontologists (who needed to explain fossil distributions) and geomorphologists (who needed to preserve the ancient verities of continental and ocean permanence). The first systematic exposition of the task before them is a letter from Schuchert to Willis, 14 April 1931:

Regarding the permanency of ocean basins, I hardly know what stand I will finally take—I am feeling my way by making others answer my questions (Barrell, and now Holmes) and face my difficulties. My difficulties you do not know and cannot appreciate because you are not a paleontologist, but you can truthfully reply, "Schuchert, you are no geologist," and this I must admit when it comes to interpreting the broader geologic structure of the earth. However, let me put up to you some of my difficulties in interpreting faunal connections.

You appear to be willing to admit a land out in the Indian Ocean going from India southwest across various islands to Madagascar. This land is Lemuria, named by Sclater long ago, and I believe in it, not so much to explain the distribution of the land animals of the Mesozoic and present, as because its presence is necessary to explain the distribution of the marine Mesozoic faunas. This Lemuria [has] . . . gone into the depths of the Indian Ocean, and what remains above sea-level is the islands you mention in your letter to me. To me these are well-established facts on the basis of the distribution of the ancient and living faunas, and can not be gainsaid. . . .

If the oceans have swallowed great marginal parts of the continents, Lemuria, and tremendous areas between Indo-China and Australia and New Zealand, why not an Atlantic bridge? The magnitude of the job is no greater than in the cases admitted and just mentioned.

Remember, I am not raising these lands—they are primordial—I am only pointing out their existence and seeking among you geomorphologists a way to get rid of them. It is up to you people to point out how it can be—or was—done, and it is up to me to prove that the faunal connections as I see them do actually exist.

Let me assure you that there is no quarrel on between us, only we must help one another. Barrell used to try hard to do so, and so does Holmes nowadays, but fails as yet to find the *modus operandi*. Will the great African explorer [Willis] find a means? Surely he will try! (WP)

Willis replied cordially, accepting Schuchert's challenge to adapt his permanence doctrine to Schuchert's necessities. The next revealing strategy discussion is again in a Schuchert to Willis letter of 1 June 1931:

I have had your very kind letter of April 21 now for over a month and have read it several times, and the more I peruse it the more I like it, and now I want to exclaim "Welcome, Brother-in-Arms, we either live or go down together!"

It is your Permian map that especially appeals to me, and particularly the paleogeography of the tropical belt. The Atlantic bridge with its peninsulas I will want to widen out some, but not more than you will allow zoogeographers. It is this admission of yours that I hail with special delight, and see one warring geologist entered into the paleontologist camp under a flag of truce. We accept you and assure you of our peace! Can I induce you to publish this map with my alterations, along with an extension of your letter to me, in the *American Journal of Science*? If so, I will take up a study of the Permian paleogeography and submit my additions and alterations to you for approval. This, then, would show that you and I agree on the main features and that you intend to defend the physical aspects of the problem, while I will do likewise regarding the paleogeography. This will give others a basement on which to build their own temples, or otherwise build them on sands.

With greetings to my recently discovered co-warrior, I am
Yours truly,
Charles Schuchert (WP)

Willis responds to Schuchert in a letter dated 9 June 1931:

I will be delighted to cooperate with you in discussing Permian paleogeography in the AJS and I await with interest your draft of the map.

My contribution would be limited, as I understand your suggestion, to the *physical basis of permanency*, which, of course, would necessarily cover the conditions of variation in the extent of lands and seas in general. . . .

As to the widening of this or that land-bridge the limiting factors admit of considerable range. Just for the sake of consistency I have assumed some bathymetric contour, say 4000 meters, and have drawn the isthmian shore lines accordingly. In general they seem adequate to connect the continental masses and to allow land enough for migrations and breeding en route. But if you need more elbow room, I will not object, always provided that the principle of the reconstruction be respected, that is; *Once a deep, always a deep. Once a continent, always a continent.* . . .

In thus revealing the depths of my heterodoxy I am not attempting to convert you from the more catholic faith of geology. Had I more of the missionary, militant spirit I would have been too busy seeking converts to have done much thinking. But since you welcome me so cordially as a "Brother in Arms" I want you to know just how hot an ember it is that you have snatched from burning. . . . So holding myself in readiness to take up the cudgels in behalf of facts rather than of the many fancies that are now so prevalent

I am cordially and happily your "co-warrior"

BW (WP; emphasis in the original)

So Willis was snatched from burning, both warriors had taken up the cudgels, and "facts" were going to triumph over fancies. Neither man realized how long it would take to work out all their differences. They coordinated their respective battle plans until near the end of 1932; by then the *AJS* had lost interest in their project, or they had decided to seek publication in the "official" geological organ, the *GSAB Bulletin.*

By early May, Schuchert had finished his paper, and sent a carbon to Willis for final comments. In it was this plea:

May I ask once more to add to your paper—the pages on Gondwana—an explanation of how a granitic Gondwana—a continental Gondwana—can be gotten rid of. Barrell tried this, and you and Lawson say he failed. . . .

Probably my article should precede yours, since I am presenting the evidence as to why we paleogeographers need land bridges, which should be followed by your discussion as to how such links are probably made and unmade.

Then let us bind together under one cover, with a special title, our two papers for personal distribution. As we need not send reprints to Fellows of the Society, probably 200 copies will suffice. I will send out these extras and tell you to whom I send them. You can then furnish me with your list and the remaining unused copies we will divide among ourselves (WP, Schuchert to Willis, 2 May 1932).

There was further dickering and editing. Willis wanted six changes in Schuchert's paper, which were accepted. Schuchert also had a list of changes, and Willis agreed. By May 12 the deed was basically done. Schuchert, the more evangelical of the two, closed his letter of that date, "This, then, I think, will close our brotherly affair, and I hope the geologists of the world will agree to a man to accept your views as to how transoceanic bridges may be made and unmade; not the bridges asked for by biogeographers crossing all parts of the oceans but the reasonable ones asked for by paleontologists! Ha! Ha! Ha! We live or die together, nicht wahr?" (WP, Schuchert to Willis, 12 May 1932).

Thus in the *GSA Bulletin* of December 1932 Schuchert's "Gondwana Land Bridges" and Willis's "Isthmian Links" occupied pages 876 to 952. Both were greatly pleased with the journal's appearance, but as usual Schuchert was the more ecstatic:

Our "grand splash," done in the colors of the rainbow, is out, and my 240 extras have arrived. Our baby is sizable and looks strong and healthy—may the geological fraternity give it the attention it deserves! . . .

Here two past-presidents appear in unison and in colors! We introduce color in our *Bulletin* and herald the news that our Society now has the cash to do the correct thing. You are responsible for the color and I am glad to share the brilliancy with you (WP, Schuchert to Willis, 30 December 1932).

The publication was spectacular; it was probably also influential. The mobilists are dismissed as inconsequential with this slight mention, "Those who lean toward the Wegener theory of continental drift as a possible explanation for the present fossil distribution discussed in this study are referred to Hoffman . . . who rejects the theory . . ." (Schuchert 1932, 878). Permanence is asserted time and again. Land bridge sinking is made to sound as easy as taking candy from a baby.

For the rest of the decade, from 1933 through 1940, the major American geological periodicals ignored Wegener and continental drift. *AJS*, which had such extensive early coverage, published only two minor items during this period. The establishment house organ *GSA Bulletin*, had only one article. The *Journal of Geology* carried nothing (no single index listing for Wegener or continental drift).

H. Frankel in 1981 found that the Schuchert-Willis *demarche* presented improved solutions to the problem of distribution of life forms "which greatly improved the status of permanentism, especially among American biogeographers and paleontologists" (p. 231). Influential the collaboration may have been, but that it was an improved solution to the problem is arguable. Hallam is witness to the isostatic difficulties that remained for the slimmed-down "isthmian links" (1972, 57).

What was the improvement? Here we had two elderly, headstrong mandarins responding to an upstart challenge from decadent Europe to hallowed American doctrines. They presented no new research, and ignored all but two arenas of the permanentist-mobilist conflict: life forms and glaciation. In this restricted arena, each compromised his long-standing private beliefs to accommodate a "brother-in-arms" in a rhetorical battle that obsessed both. The texts of the two articles, when read critically, reveal the extent to which Willis glossed over isostatic problems by relabeling land bridges as isthmian connections, thus satisfying Schuchert's need to sink continental terrains; and Schuchert gave in to Willis's *idee fixe* by agreeing that oceanic basins had always been oceanic. This is neither science nor scholarship.

Both men remained dogmatic about mobilism to the bitter end. D. Griggs recalled in 1973 his exposure to Willis's intransigence. Griggs was responding to the presentation of the Author L. Day medal:

> Dr. Day was president of the GSA [1938] when I gave my paper on convection currents in the mantle, complete with a working model and a movie which showed the lithosphere plunging under the continents creating massive subduction, meanwhile sweeping the sedimentary layer from the oceans. Harry Hess presided, and endeavored to get favorable discussion of these then controversial ideas, but circumstances prevented him. Andy Lawson, sitting in the front row got up and squeaked, "I may be gullible, I may be gullible! But I'm not gullible enough to swallow this poppy-cock." After his long tirade, before Harry could do anything, Bailey Willis who was sitting in the second row got up, turned to face the audience and said, "All you here today bear witness—for the first time in twenty years, I find myself in complete agreement with Andy Lawson." (1974, 1342)

Was It Rational?

One need not enter the convoluted discussions among philosophers about rationality to justify marketplace evaluations of the positions and rhetoric of the American participants in these debates. In this 1920–1932 period, the dark side of American chauvinism and recalcitrance was evident.

The major sin of the zealots was their rigidity and refusal to consider any proposal for mobilism. They need not have accepted Wegener's flawed sial rafts plowing through a tougher sima; nor need they accept the questionable Greenland measurements. But their bitter attempts to make Wegener out a fool, or a dreamer, or a poet, were nothing but intransigence. They were guilty of what Dott (1981) calls the "single-model syndrome," and their culpability is greater because they preached open-mindedness and multiple working hypotheses.

The mobilists, by contrast, were less fanatical and recognized that much more research would be needed to fully understand how the continents moved. Even Wegener acknowledged that alternative hypotheses of displacement existed. As Jacoby (1981) points out, Wegener had early espoused a form of seafloor spreading, but tended to ignore it in his later books. However, even in the third edition, Wegener stated about Madagascar, "[It] is possible that it has been passively carried along by a stream of sima" (1924, 191). And he clearly was nondogmatic about displacing forces, "The Newton of Drift theory has not yet appeared" (1966, 167).

Partly the intransigence of these American opponents of mobilism was due to their absorption in the local diastrophism at their doors—no massive overthrusts as in the Alps, no emerging mountains and oceanic deeps as in the East Indies, no startling petrographical and faunal resemblances like those that so impressed du Toit. But this does not excuse them from their disregard of the data cited and the arguments made by the convinced, though not fanatical, mobilists elsewhere: Holmes, Bailey, Brouwer, van der Gracht, Collet, Argand, du Toit and others. Parochialism is understandable; it can be carried too far. Mobilists abroad did *not* sin by ignoring the best American findings.

Nor was the American put-down of Wegener because he was not a geologist rational. A sophisticated concept of expertise would have been useful here. Of course when one seeks to learn whether the African-Brazilian faunal or petrographical similarities are significant one must find an expert—preferably with extensive experience on both continents, and necessarily one with an open mind.

Several writers have commented that precisely because Wegener was not an expert in paleontology or geology proper his conceptual facilities could escape the parochialism of those fields and embrace a unifying hypothesis that made sense for all the earth sciences (Hallam 1973, 113; Macrakis 1984, 166). The elder Chamberlin, especially, should have appreciated this ability, because early in his career he jumped the boundaries of glaciology to end up with a cosmogony that was clearly "outside" his field. Mobilism was rejected in the 1920s, certainly in America and to some extent elsewhere, because Wegener was not a member of the fraternity.

Oreskes (1988, 339) may be right in arguing that early mobilist hypotheses were destined for rejection because the evidence for them (paleontological, petrographical, geographical, and paleoclimatic) was not hard, observer-independent data such as paleomagnetic measurements. Her claim that Americans were *less* impressed by soft data then geologists elsewhere is suspect. The leaders of the crusade against mobilism used observer-dependent data extensively.

Dott's speech to the 1981 AAAS meeting in Toronto gives a comprehensive explanation of "The North American Rejection of Wegener's Drift Theory." The ruling hypothesis of continental accretion (Dana) seemed to explain North America well enough. The major proponent of drift, Wegener, was foreign in specialty as well as nationality. American geologists were more provincial than Europeans, lacking national involvement in far-flung colonial possessions. Fixed continents seemed more consistent with Lyellian uniformitarianism, whereas Wegener's emphasis on one major drift episode smacked of catastrophism. Geophysicists, led by H. Jeffreys, disdained any possibility of a force sufficient to move continents. And American geologists were simply very conservative.

But one major factor remains to be considered: the vigorous campaign *against* drift carried out by prestigious figures at the top of the geological hierarchy. To the lone elite mobilist in the United States what was happening was clear and invidious. Daly expressed himself this way to Willis when Willis asked him to participate in the continental problems panel at the 1928 GSA meeting (WP, Daly to Willis, 8 September 1928):

My dear Willis,
. . . . Your subject has been very much on my mind for two decades or more, ever since I opened the first course ever given at Harvard on Oceanography.
    If the symposiasts can throw new light on this difficult problem they will make a ten-strike. You make light of horizontal movements of the Sial, but I see no way of dodging this idea *if* the symposiasts are really going to clear ground for us. You say that my entertaining the possibility of Continental Drift (and as a *possibility* only do I entertain it, and probably only as such will I ever entertain it, so far as migration of thousands of Km is concerned) is possibly my "undoing". But I am in a rattling good company of liberals in thought and fear not the slightest any "undoing" because of open-mindedness

for a European idea. Is the symposium to ignore European thought entirely? How else can it ignore the migration theory? Mind you, I am not writing of Wegener; I am thinking of *Willis's* (every American geologist's) responsibility in this matter. I believe that it is *your* bounden duty and the bounden duty of every one of your symposiasts to get really busy on the general idea of horizontal migration of continental blocks—and let Wegener alone in the process!! And me too!!! Go to it with an open mind and see if you can beat us all at putting the idea in better shape. . . .

I hope the symposiasts will attack the origin of the sial—unescapable in fact if one is to talk ocean-basin history. Very few of us have tackled this vast problem. Chamberlin's speculations seem to me useless here; the moon-scar theory is pretty hopeless (as voiced by Pickering etc.); it is high time for someone else to endanger his soul and make a new guess—or a dozen guesses. Perhaps in that way we shall "bump our heads against the truth."

Ever sincerely yours,
Reginald A. Daly

Daly did not live to see the acceptance of plate tectonics in 1967.

In his comment on J. Dewey's review of Argand, R. Scholten identifies precisely how the advances of the European mobilists were anathematized in America, "The opinion makers were permitted to stop the tide. The grand synthesis fell under the tyranny of a ruling theory pronounced as 'the best' by those we chose to think of as 'the best' " (1978, 586).

In a letter to the author, Scholten worries that things may not have changed since the 1920s and 1930s, "Are heretical proposals (for funding and publication) received more kindly and with a greater humility today?" (Scholten to Newman, 28 August 1992). The same worry is expressed in P.D. Lowman Jr.'s lead article of *New Concepts in Global Tectonics:*

The theory [plate tectonics] is a useful pedagogical tool in being coherent, simple, visual, and easily taught. However, these strengths are outweighed by weaknesses: deceptive simplicity, limited scope, and possibly fundamental incorrectness. These problems are already slowing geologic progress by inducing narrow, rigid viewpoints, and discouraging research by convincing students that the main problems are solved. (1992, 3)

Lowman then develops a number of ways in which the stifling authoritarianism of the 1930s and 1940s can be avoided. As one reads *New Concepts,* one is convinced that orthodoxy has not suppressed alternative hypotheses. One hopes it has not threatened the livelihood of heretical theorists. The unsavory polemics of the early drift debate seem to be absent. Geology has, again, multiple working hypotheses.

NOTES

Thanks are due for helpful comments to Robert H. Dott, Jr.; Henry Krips; Ted McGuire; Trevor Melia; Walter Pilant; Anthony Prave; John E. Sanders; Robert Scholten; and Curt Teichert. For indispensable translations of the great European writers on tectonics, Albert and Marguerite Carozzi have my heartfelt thanks. Passages and letters from the Bailey Willis Collection are reproduced by permission of The Huntington Library, San Marino, California.

1. A thorough discussion of the lengthy development of the planetesimal hypothesis is in Brush (1978). Chamberlin presented his hypothesis in an extensive series of articles in the *Journal of Geology* from 1898 through 1927. A condensation of his thought appeared in the Carnegie Institution of Washington yearbooks, 1904 through 1928. M. Greene (1982) places Chamberlin in the context of European and American attempts at geological synthesis.

REFERENCES

Boone, A. W. 17 November 1926. "Stars to Show if the Continents Move." *New York Times:* 4.
Bowie, W. 1925a. "Isostasy at Madrid Meeting of International Geophysical Union." *Pan-American Geologist 43:* 337–44.
———. 1925b. "Miscellaneous Scientific Intelligence." *American Journal of Science 9:* 180–85.
———. 6 September 1925c. "Scientists to Test 'Drift' of Continents." *New York Times:* 5.
Brouwer, M. A. 1921. "The Horizontal Movement of Geanticlines and the Fractures Near Their Surface." *Journal of Geology 29:* 560–77.
———. 1926. "Architecture of Dutch East Indies." *Pan-American Geologist 45:* 201–10.
Brush, S. G. 1978. "A Geologist Among Astronomers: The Rise and Fall of the Chamberlin-Moulton Cosmogony." *Journal of the History of Astronomy 9:* 1–41, 77–104.
Carnegie Institution of Washington Yearbooks. 1904–1928. Washington, D.C.
Carozzi, A. V. 1977. "Editor's Introduction." In E. Argand, ed., *Tectonics of Asia.* New York: MacMillan, pp. xiii–xxvi.
Chamberlin Papers. Department of Special Collections. University of Chicago Library.
Chamberlin, R. C. 1928. "Some of the Objections to Wegener's Theory." In *Theory of Continental Drift.* Tulsa, Okl.: American Association of Petroleum Geologists, pp. 83–87.
Chamberlin, T. C. 1890. "The Method of Multiple Working Hypotheses." *Science 15:* 92–96.

———. 1923. "An Ancient Theory of the Pacific." *Journal of Geology* 31: 152–53.

Collet, L. W. 1926. "The Alps and Wegener's Theory." *Geographical Journal* 67: 301–12.

Daly, R. A. 1923. "The Earth's Crust and Its Stability." *American Journal of Science* 5: 24–371.

Dewey, J. F. 1978. "La Tectonique de L'Asie (Tectonics of Asia)." *Geology* 6: 190–91.

Diener, C. [1915] 1922. "Major Features of Earth's Surface." *Pan-American Geologist* 37: 177–97.

Dott, R. H., Jr. 1981. "The North American Rejection of Wegener's Drift Theory." American Association for the Advancement of Science, Toronto Meeting.

———. 1992. "T. C. Chamberlin's Hypothesis of Diastrophic Control of Worldwide Changes of Sea Level." In R. H. Dott, Jr., ed., *Eustasy: The Historical Ups and Downs of a Major Geological Concept.* Boulder: Geological Society of America, Memoir 180.

"Drifting Apart." 6 September 1925. (Editorial). *New York Times:* 6.

du Toit, A. L., and R. C. Reid. 1927. *A Geological Comparison of South America with South Africa.* Washington, D.C.: Carnegie Institution of Washington.

Frankel, H. 1981. "The Paleobiographical Debate Over the Problem of Disjunctively Distributed Life Forms." *Studies in the History and Philosophy of Science* 12: 211–59.

Friedman, G. M., and J. E. Sanders. 1978. *Sedimentary Strata and Tectonic Movements.* New York: Wiley.

Greene, M. T. 1982. *Geology in the Nineteenth Century.* Ithaca: Cornell University Press.

Griggs, D. T. 1974. "Response by David Griggs." *Geological Society of America Bulletin* 85: 1342–43.

Hallam, A. 1972. "Continental Drift and the Fossil Record." *Scientific American* 227: 57–66.

———. 1973. *A Revolution in the Earth Sciences.* Oxford: Clarendon Press.

Holmes, A. 1928. "Continental Drift." *Nature* 122: 431–33.

Holtedahl, O. 1920. "Paleogeography and Diastrophism in the Atlantic-Arctic Region During Paleozoic Time." *American Journal of Science* 49: 1–25.

Ingalls, A. G. 1926. "Are the Continents Drifting?" *Scientific American* 134: 8–9.

Jacoby, W. R. 1981. "Modern Concepts of Earth Dynamics Anticipated by Alfred Wegener in 1912." *Geology* 9: 25–27.

Keith, A. 1923. "Outlines of Appalachian Structure." *Geological Society of America Bulletin* 34: 310–76.

Laudan, R. 1985. "Frank Taylor's Theory of Continental Drift." *Earth Science History* 4: 118–21.

LeGrand, H. E. 1988. *Drifting Continents and Shifting Theories.* Cambridge, England: Cambridge University Press.

Leith, C. K. 1922. "Composite Nature of Rock Mass-Movement," *Pan-American Geologist 37:* 84–86.

Lowman, P. D., Jr. 1992. "Plate Tectonics and Continental Drift in Geologic Education." In S. Chatterjee and N. Hotton, Ill, eds., *New Concepts in Global Tectonics.* Lubbock: Texas Tech University, pp. 3–12.

Macrakis, K. 1984. "Algred Wegener: Self-Proclaimed Scientific Revolutionary." *Archives Internationales d'Histoire des Sciences 112:* 163–76.

Marvin, U. B. 1973. *Continental Drift.* Washington, D.C.: Smithsonian Press.

Melton, F. A. 1930. "Theory of Continental Drift—A Symposium." *Journal of Geology 38:* 284–86.

Oreskes, N. 1988. "The Rejection of Continental Drift." *Historical Studies in the Physical and Biological Sciences 18:* 311–48.

"Proceedings: Washington Academy of Sciences, 179th Meeting." 1923. *Journal of the Washington Academy of Sciences 13:* 445–50.

Richarz, S. 1926. "Climates of the Geologic Past." *Pan-American Geologist 45:* 97–99.

Scholten, R. 1978. "Comment on Tectonics of Asia, by E. Argand." *Geology 6:* 585–86.

———. 28 August 1992. Letter to Newman.

Schuchert, C. 1928. "The Hypothesis of Continental Displacement." In *Theory of Continental Drift.* Tulsa, Okl.: American Association of Petroleum Geologists, pp. 104–44.

———. 1932. "Gondwana Land Bridges." *Geological Society of America Bulletin 43:* 876–915.

Taylor, F. B. 1910. "Bearing of the Tertiary Mountain Belt on the Origin of the Earth's Plan." *Geological Society of America Bulletin 21:* 179–226.

Termier, P. 1925. "The Drifting of the Continents." *Annual Report for 1924.* Washington, D.C.: Government Printing Office.

Thornburg, D. H. 1925. "The Origin of Continents and Oceans, by Alfred Wegener." *American Association of Petroleum Geologists Bulletin 9:* 916–18.

van der Gracht, W. 1926. "Discussion of Continental Drift at Meeting, November 15–17." *American Association of Petroleum Geologists Bulletin 10:* 1002–03.

———. 1928. *Theory of Continental Drift.* Tulsa, Okl.: American Association of Petroleum Geologists.

Wegener, A. L. 1924. *The Origin of Continents and Oceans.* 3rd ed. Translated by J. G. A. Skerl. London: Methuen.

———. 1966. *The Origin of Continents and Oceans.* 4th ed. Translated by J. Biram. New York: Dover.

"Wegener's Displacement Theory." 1922. *Nature 109:* 202–03.

Willis, B. 1927. "Folding or Shearing, Which?" *American Association of Petroleum Geologists Bulletin 11:* 31–47.

———. 1928a. "Continental Drift." In *Theory of Continental Drift.* Tulsa, Okl.: American Association of Petroleum Geologists, pp. 76–82.

———. 1928b. "Continental Genesis." *Geological Society of America Bulletin 40:* 281–336.

————. 1928c. "Growing Mountains." *Scientific American 138:* 151–53.

————. 1932. "Isthmian Links." *Geological Society of America Bulletin 43:* 917–52.

Willis Papers. Huntington Library, San Marino, California.

Wood, R. M. 1985. *The Dark Side of the Earth.* London: Allen & Unwin.

9

Topics, Tropes, and Tradition

Darwin's Reinvention and Subversion of the Argument to Design

John Angus Campbell
Department of Speech Communication, University of Washington

On the Origin of Species (Darwin 1967) is a work at once accessible and strange. It speaks so clearly the idiom of the layperson whether of the nineteenth century, or, with certain allowances, of today. The table of contents displays the book's thematic progression so clearly as to tell its own story, running from "Variation Under Domestication," to "Variation Under Nature," to "Struggle for Existence," to "Natural Selection" without a break. The introduction, "When on board H.M.S. 'Beagle,' as naturalist, I was much struck with certain facts . . ." charms the reader as it tells of the apparently chance beginning of the present volume and apologizes for its incomplete form. The first line of the first chapter draws the reader into the company of an author who welcomes the reader as his companion on the second voyage of intellectual discovery.

When we look to the individuals of the same variety or subvariety of our older cultivated plants and animals, one of the first points which strikes us, is, that they generally differ much more from each other, than do the individuals of any one species or variety in a state of nature. When we reflect on the vast diversity of the plants and animals which have been cultivated, and which have varied during all ages under the most different climates and treatment, I think we are driven to conclude that this greater variability is simply due to our domestic productions having been raised under conditions of life not so uniform as, and somewhat different from, those to which the partent-species have been exposed. (Ibid., 7)

The facts may be new or unfamiliar, the sentences sometimes lengthy, yet the reasoning seems clear, engaging and almost easy. When we examine these first two sentences together we find the following sequence of clauses: "When we look . . . when we reflect . . . I think we are driven to conclude." Readers know in encountering this language they are being led, and may wish to follow—or may try to resist. In either event, readers soon discover The *Origin* is not a book in which they may easily be spectators. The Darwin reader is constantly edged into becoming a participant, indeed (if I may borrow an expression from the Watergate years) "an unindicted co-conspirator." Some passages make it difficult for readers to avoid being implicated. Readers cannot easily resist sentences such as "He must be a dull man who can examine the exquisite structure of a comb, so beautifully adapted is its end, without enthusiastic admiration" (ibid., 224), or the Darwin children's favorite passage on larval barnacles, "In the second stage, answering the chrysalis stage of butterflies, they have six pairs of beautifully constructed natatory legs, a pair of magnificent compound eyes, and extremely complex antennae" (ibid., 411). F. Darwin (1911, 114) records how he and his siblings compared this description to an advertisement.

Yet the *Origin* is also strange. In the midst of what in principle are highly technical chapters, such as "The Struggle for Existence" (chap. 3) in which Darwin introduces the core of his theoretic insight—the application of Malthus's law of population dynamics to the unstaunchable flow of organic variation—readers may feel as though they were reading curious facts from a popular magazine of natural history:

I am tempted to give one more instance showing how plants and animals, most remote in the scale of nature, are bound together by a web of complex relations. . . . I have . . . reason to believe that humble bees are indispensable to the fertilisation of the heartsease (Viola tricolor), for other bees do not visit this flower. From experiments which I have tried, I have found that the visits of bees, if not indispensable, are at least highly beneficial to the fertilisation of our clovers; but humble-bees alone visit the common red clover (Trifolium pratense), as other bees cannot reach the nectar. Hence I have very little doubt that if the whole genus of humble-bees became extinct or very rare in England, the heartsease and red clover would become very rare, or wholly disappear.

The number of humble-bees in any district depends in a great degree on the number of field-mice, which destroy their combs and nests; and Mr. H. Newman, who has long attended to the habits of humble-bees, believes that "more than two thirds of them are thus destroyed all over England." Now the number of mice is largely dependent, as everyone knows, on the number of cats; and Mr. Newman says, "Near villages and small towns I have found the nests of humble-bees more numerous than elsewhere, which I attribute to the number of cats that destroy the mice." Hence it is quite credible that the presence of a feline animal in large numbers in a district might determine, through the intervention first of mice and then of bees, the frequency of certain flowers in that district! (Darwin 1967, 73–74)

That the delight a present-day reader may take from this passage was shared by Darwin's contemporaries is indicated by the reluctant testimony of an American critic, "The theory is certainly ingenious and captivating; and . . . the whole exposition of it very delightful" (Bowen 1860, 481). But this praise sports a sharp undertone, for the reviewer then adds a qualification which has also been echoed by Darwin readers to the present day, "It seems almost ungrateful, in return for the entertainment which the work has afforded, to . . . weigh the evidence and ascertain. . . . If it is true" (ibid., 481–483). While we may dismiss this and other critics as reactionaries incapable of weighing the evidence, we may be more prudent to let this and similar statements stand as examples of the offenses that the *Origin* gave to its immediate audience and which it is capable of giving still. Darwin taught that adaptation in nature was never perfect; we may apply this same principle to audience adaptation, not excluding Darwin's contemporaries. Indeed, if we take S. Gould's contrast between Darwinism and the path Gould believes "a wise Creator" (1983, 258) would have followed, we gain a certain perspective by incongruity into some of the most rhetorically interesting features of Darwin's achievement. In Darwin's world, Gould explains things are "jerry-built of available parts used by ancestors for other purposes" (1977, 91) and that "a perfect engineer would certainly have come up with something better. . . . The best illustration of adaptation by evolution are ones that strike our intuition as peculiar or *bizarre*" (ibid.). Or again, "Odd arrangements and funny solutions are the proof of evolution—paths that a sensible God would never tread but that a natural process, constrained by history, follows perforce" (Gould 1980, 20–21).

If we take the attributes Gould has claimed for a "sensible God" and turn them back on Darwin, we have a picture of how Darwin might have expressed himself had he been a "sensible genius"—one presumably dedicated to the purity of theory over success, contemptuous of audience expectations or the proximate calculation of trajectories for falling chips, and not constrained by the legacy of previous genres. Had Darwin been such a person he might have expressed, as Gould has, the central thesis of Darwinism:

> 1. Organisms vary, and these variations are inherited (at least in part) by their offspring.
>
> 2. Organisms produce more offspring than can possibly survive.
>
> 3. On the average, offspring that vary most strongly in directions favored by the environment will survive and propagate. Favorable variation will therefore accumulate in populations by natural selection. (Gould 1977, 11)

Gould's version, similar to accounts likely to be found in other short summaries, sensibly puts Malthus's laws of population in the foreground and downplays the metaphor of selection. Indeed "natural selection," as a metaphor, is dead on arrival in this account. For all the work the metaphor, or personification, "selection" does in this version, Darwin might have avoided a great deal of grief and simply called his theory something innocuous like "the theory of natural remainders." (In private letters Darwin suggested that were he to rewrite his book he would have used the term "natural preservation." See letter to W. H. Harvey, August 1860 [Darwin 1903, vol. 1, 160–61]; letter to Lyell, 6 June 1860 [Darwin 1911, vol. 2, 111].) One imagines Gould's version of Darwin's central tenets being uttered by a proverbial white-coated figure speaking in appropriately impersonal tones. Consider how differently, indeed how strange, this "sensible idea" sounds in Darwin's language. Consider how difficult it is to read without hearing echoes (either despite or because of the content) of a black-robed figure with a round white collar and a stained-glassed voice:

As man can produce and certainly has produced a great result by his methodical and unconscious means of selection, what might not nature effect? Man can act only on external and visible characters: nature cares nothing for appearances, except insofar as they may be useful to any being. She can act on the whole machinery of life. Man selects only for his own good; Nature only for that of the being which she tends. Every selected character is fully exercised by her, and the being is placed under well-suited conditions of life. Man keeps the natives of many climates in the same country, he seldom exercises each selected character in some peculiar and fitting manner; he feeds a long- and a short-beaked pigeon on the same food; he does not exercise a long-backed or long-legged quadruped in any peculiar manner; he exposes sheep with long and short wool to the same climate. He does not allow the most vigorous males to struggle for the females. He does not rigidly destroy all inferior animals, but protects during each varying season, as far as lies in his power, all his productions. He often begins his selection by some half-monstrous form; or at least by some modification prominent enough to catch his eye, or to be plainly useful to him. Under nature, the slightest difference of structure or constitution may well turn the nicely-balanced scale in the struggle for life, and so be preserved. How fleeting are the wishes and efforts of man! How short his time! and consequently how poor will his productions be, compared with those accumulated by nature during whole geological periods. Can we wonder, then, that nature's productions should be far 'truer' in character than man's productions; that they should be infinitely better adapted to the most complex conditions of life, and should plainly bear the stamp of far higher workmanship?

It may be said that natural selection is daily and hourly scrutinising, throughout the whole world, every variation, even the slightest; rejecting that which is bad, preserving and adding up all that is good; silently and insensibly working wherever and whenever opportunity offers, at the improvement of each organic being in relation to its organic and inorganic conditions of life. We see nothing of these slow changes in progress, until the hand of time has marked the long lapses of ages, and then so imperfect is our view into long past geological ages, that we only see that the forms of life are now different from what they formerly were. (Darwin 1967, 83–84)

What are we to say of a passage that in form is an exhortation, that echoes of the *Psalms* (39:5, 139:5–14), and reverberates with the very accents of the worldview Darwin is so celebrated for replacing? Indeed one may ask, as did Darwin's colleagues and contemporaries, whether a semideity, some special cosmic force, was doing the selecting. Nor is this the only instance in which Darwin has taken a concept central to his theory (which in other contexts—his notes or in later less popular works—he was able to express nearly as succinctly as has Gould) and cast it in language that, for all its semantic range and rich rhetorical suggestiveness, has left the meaning imprecise, even misleading.

In a typically perceptive essay M. Hodge, drawing on the work of V. Kovalowski, has shown Darwin's use in the early drafts of his work of the *vera causa* principle. This principle, as cast in its modern form by Newton and passed down the stream of tradition from Reid to two thinkers influential on Darwin, J. Herschel and C. Lyell, became the expository skeleton around which Darwin structured his 1842 "sketch" of his thesis, also his 1844 "essay," his abandoned manuscript *Natural Selection,* and even his classic *Origin of Species.* According to this principle, to show that something qualifies as a cause one must establish its existence (independent of the phenomena one wishes to account for) and, as a separate issue, its competence to account for the change one wishes to explain. In Darwin's early manuscripts this principle takes on a tripartite form of establishing the existence, the competence, and the responsibility of variation and selection for effecting organic change. The problem Hodge finds in Darwin's exposition is that the more Darwin wrote, the more obscured became his organizing principle. "[In] writing the *Essay, Natural Selection* and the *Origin,* Darwin made successive departures from his early version of the structure, and that, though really minor and superficial, these departures were eventually enough to render the strategy and organization of his most famous book unhelpfully and quite unnecessarily obscure" (1977, 242).

Figure 9.1 from Hodge's essay shows the basic structure of Darwin's 1842 essay as the latter has reconstructed it. As we can see from the list of chapters, the *Origin* does not have the clear thematic divisions that we find in the drafts:

Chapters of the Origin

1. Variation Under Domestication
2. Variation Under Nature
3. Struggle for Existence
4. Natural Selection
5. Laws of Variation
6. Difficulties of the Theory
7. Instinct
8. Hybridism
9. On the Imperfection of the Geological Record
10. On the Geological Succession of Organic Beings
11. Geographical Distribution
12. Geographical Distribution (continued)
13. Mutual Affinities of Organic Beings: Morphology: Embryology: Rudimentary Organs
14. Recapitulation and Conclusion

When we look at the *Origin* as it appears we see the initial suite of four chapters, then an apparent miscellany of chapters on various subjects and objections. This is followed by a suite of chapters on geology and geographic distribution and a final chapter on embryology and mor-

FIGURE 9.1   The Structure of the *Sketch* of 1842

Part I	Topic a	Chapter I		Consideration 1		Division One
Principles of variation and selection under domestication	b c	Variation under domestication		Existence case		Natural selection established as VCP cause for species
Part II	a	Chapter II				
Application of those principles to species under nature	b c d e f	Variation in nature, and natural selection		Cons. 2	The case made	
				Competence case	Difficulties considered	
	g	Chapter III	Instinct etc.			
Part III		Chapter IV	Geology	Consideration 3		Division Two
Trial of theory of natural selection as explanatory of species production		Chapter V	Geology	Responsibility case		Natural selection as on balance, probably responsible for species
		Chapter VI	Geography			
		Chapter VII	Classification			
		Chapter VIII	Morphology			
		Chapter IX	Morphology			
		Chapter X	Recapitulation			

phology before Darwin's recapitulation. Hodge offers good evidence that the *Origin* is in fact better organized than it seems, as illustrated in his diagram in figure 9.2.

Indeed, Hodge shows that Darwin adhered to this very same logic in what was to have been his opus *Natural Selection*—the immediate predecessor of the *Origin*. See figure 9.3. Hodge concludes by contrasting the lack of a clear *vera causa* sequence in the *Origin* with the exceedingly clear execution of that same principle in the opus of Darwin's mentor, Lyell, "But Darwin, it hardly needs remarking, had little of that love of schematic clarity and consistency which kept Lyell to

FIGURE 9.2 The Structure of the *Origin* (1859)

Part I	Chapter I	Consideration 1		Division One
Variation and selection under domestication		Existence case		Natural selection established as VCP cause for species
Part II	II			
Variation and selection under nature	III			
	IV	Consideration 2	The case	
	V	Competence case	Difficulties considered	
	VI			
	VII			
	VIII			
Part III	IX	Consideration 3	Geological difficulty	Division Two
Trial of theory of natural selection as explanatory of species production	X	Responsibility case	Evidence favouring responsibility	Natural selection as probably responsible for species production
	XI			
	XII			
	XIII			
Recapitulation	XIV			

the same format through no less than twelve editions of the *Principles* and more than forty years of revisions, some of them substantial— even radical" (Hodge 1977, 245). In the end, however, Hodge does not see Darwin's organization as a handicap, "not at the time anyway, for the cognoscenti would have had little trouble recognizing the basic organization of the *Origin*, even obscured as it was by Darwin's mingling of arguments and shuffling of topics that he had once had explicitly, though privately, arranged in their natural order. It does, however, make it easy for his readers today to miss the wood for the trees" (ibid.).

My concern is less with the clarity, or lack of it, in the logical arrangement of the chapters of the *Origin* and more with Hodge's point that Darwin's arrangement has unnecessarily obscured the true logic of his argument. I argue that the *Origin* is arranged as it is, partially

FIGURE 9.3   The Structure of *Natural Selection* (1856–8)

Part I Domestic variation and artificial selection	[Chapter I] [II]	[Variation under dom.] [Variation under dom.]		Division One Natural selection as VCP cause
Part II Natural variation and selection	III IV V VI VII	Crossing etc. Natural variation Struggle for existence Natural selection Laws of variation	Existence and competence cases	
	VIII IX X	Perfect organs etc. Hybrids Instinct	Difficulties for competence case	
Part III The theory tested	XI	Geographical distribution	Responsibility case	Division Two Natural selection as responsible

at least, because, as in the case of his insight into Malthus, Darwin rhetorically reconstructed the *vera causa* principle to meet the rhetorical needs of persuasive presentation.

I focus on two of Darwin's most important abstractions, his central metaphor, or personification, "natural selection" and his tactical use of the *vera causa* principle. I examine how each is a rhetorical recreation—or a translation into "prudential logic"—of what originally was an algorithmic logic. In the course of my remarks I show how "square parts" and "funny solutions," particularly adaptations to the English tradition of natural theology, were central to Darwin's attempt to use the very conventions of prior tradition to overcome tradition.

On the Way to Natural Selection: From Malthus to the Breeder

In early 1837, following his return from the Beagle voyage in September 1836, Darwin sketched his first notes on "transmutation." In June 1837, he opened the first of what became four "transmutation" notebooks lettered "B" through "E" (notebook "A" having been on geology; Barrett et al. 1987). Having gone through several "pre-Malthusian" theories Darwin, in September 1838, recorded the following crucial entry in notebook "D" (see Kohn 1980, esp. 136–54, Kohn and Hodge 1985). The important part, at least for our present purposes, of this important passage is as follows:

<<I do not doubt, every one till he thinks deeply has assumed that increase of animals exactly proportional to the number that can live.—>>We ought to be far from wondering of changes in number of species, from small changes in nature of locality. Even the energetic language of <Malthus> <<Decandoelle>> does not convey the warring of the species as inference from Malthus.—<<increase of brutes, must be prevented solely by positive checks, excepting that famine may stop desire.—>>in Nature production does not increase, whilst no checks prevail, but the positive check of famine consequently death. (134e)

population increase at geometrical ratio in FAR SHORTER time than 25 years—yet until the one sentence of Malthus no one clearly perceived the great check amongst men.—<<Even a *few* years plenty, makes population in Men increase, & an ordinary crop. causes a dearth. . . . take Europe on an average, every species must have same number killed, year with year, by hawks. by. cold & c— . . . even one species of hawk decreasing in number must affect instantaneously all the rest.—One may say there is a force like a hundred thousand wedges *trying* to force <*into*>every kind of adapted structure into the gaps<of>in the oeconomy of Nature, or rather forming gaps by thrusting out weaker ones.<<The final cause of all this wedgings, must be to sort out proper structure & adapt it to change.—to do that, for form, which Malthus shows, is the final effect, (by means however of volition) of this populousness, on the energy of Man>> (135e)

Two points are particularly noteworthy in this entry. First, this passage is a trope of thought. Darwin has found in the midst of what to Malthus was an argument for stasis a case for change. The key expression "<<The final cause of all this wedgings, must be to sort out

proper structure & adapt it to change—"at once confirms Malthus and turns him on his head. This is Malthus supplemented and transformed by Darwin's own keen sense of the unstaunchableness of variation—it is no longer Malthus pure and certainly not simple. Second, the image of "a hundred thousand wedges" portrays a force that is imminent and self-sufficient—a nature that is at once the "hundred thousand wedges," the driving and the origin of the "adapted structure[s]." As D. Kohn has hauntingly put it, this telegraphic note represents the moment when Darwin saw "the creative power of death" (1980, 144). But in the spirit of Sherlock Holmes's *Silver Blaze,* one cannot help but note the astonishing feature of this passage—the deafening silence of its nonbarking dogs. We have here the very heart of Darwin's theory without any allusion to domestic breeding. As always Darwin is metaphorical, but in his first metaphor, concept and image perfectly agree. Though the expression "adapted structure" most likely represents the last gap of Darwin's previous belief that adaptations were already perfect, a belief he will soon abandon, nothing else suggests a guiding providential force. If ever there were a biological counterpart to the mindless, directionless mechanisms of uniformitarian geology, this is it.

Darwin's first clear connection between his new theory and domestic breeding occurs on 2 December 1838, more than two months after his reading of Malthus. The passage is casual, as though Darwin has not fully realized what his comment has shown, "Are the feet of waterdogs at all more webbed than those of other dogs.—if nature had had the picking she would make <them> such a variety far more easily than man,—though *man's practiced* judgment even without time can do much.—[E63:414]

The passage provides a modest version of Darwin's later more extravagant personification of nature as breeder. Let us also note Darwin's use of the breeder's term "picking" for the more elegant "selection" and that the analogy flows from domestic animals—water dogs—to nature and then from nature back to domestication (Limoges 1970, 105). That Darwin has begun to realize the possibility of a new theory with his new analogy begins to be clearer two weeks later in an entry from December 16:

It is a beautiful part of my theory, that <<domesticated>> races of <a> organics are made by precisely same means as species—but latter far more perfectly & infinitely slower.—No domesticated animal is perfectly adapted to external conditions.—hence great variation in each birth) from man arbitrarily destroying certain forms & not others.—[E 71:416]

The organic or narrative continuity of Darwin's new theory with his preselectionist (or at this point "prepicking") theory is evident by the direction of this example. In this passage, in other words, if we understand evolution in nature first, we will see a parallel between it and what the domestic breeder does. In the remainder of the line he connects variation with taxonomy—which is the expository pattern connecting the first and second chapters of the *Origin*, "Variation Under Domestication" and "Variation Under Nature." Various additional passages suggest Darwin's rhetorical-conceptual oscillation back and forth between images which exploit the potential of his new breeding analogy and images which have greater affinity with his initial Malthusian insight into chance and the imminent self-sufficiency of natural law. If some passages suggest Darwin has found a convincing analogy, one passage suggests it was unconvincing when it was first tried. On May 4, Darwin records a note from a conversation with his sister and brother-in-law:

<<Elizabeth & Hensleigh. seemed to think it absurd that the presence of the Leopard & Tiger together depended on some nice qualifications each possess., & that tiger springing an inch further would determine his preservation—. . . then that quality which saved him, would be the one encouraged—>>[E 144 439]

The problem in this passage seems to be how to present slight variations so that they can form a credible account of organic difference. In a passage that shortly follows, Darwin seems to take over their objection and turn it into a kind of defense:

It seems absurd proposition, that every <<budding>>tree, & every buzzing insect & grazing animal owes its form, to that form being <<the one alone>> out of innumerable other ones, <alone> <<which has been>> preserved.—but be it remembered, how little part of the Grand Mystery is this, the law of growth, that which changes the acorn into the oak.—In short all which <<Nutrition, growth & reproduction>>is common to all living beings. vide Lamarck Vo II. p. 115 4 four laws. [E145]

Despite its complexity, the central theme of this passage is the expository problem of how to give random variation direction, and to show how systematic form could emerge from anything so apparently formless. Darwin's robust notion of variety—the numerous possibilities of form—seems to have been more powerful than his ability to explain what could shape variation and preserve structure. Hence, Darwin's Malthusian arguments, however he presented them exactly, seem to have gotten him into a hole from which he is trying to rescue himself by relying on the earlier explanatory conventions which he adopted from Lamarck.

In his "sketch" of 1842 and in his larger "essay" of 1844 all of this changed dramatically. At the core of his first exposition of his theory we find an image that is at once familiar yet strange. It is familiar in that it is so like the expository image at the heart of the *Origin,* yet strange in that it is cast in a markedly different form. In the first several pages of his "sketch" Darwin begins, as with the *Origin,* with an account of domestic breeding:

But if every part of a plant or animal was to vary . . . and if a being infinitely more sagacious than man (not an omniscient creator) during thousands and thousands of years were to select all the variations which tended towards certain ends ([or were to produce causes that tended toward the same end)], for instance, if he foresaw a canine animal would be better off, owing to the country producing more hares, if he were longer legged and keener sight—greyhound produced. If he saw that aquatic animal—skinned tones. . . . Who, seeing how plants vary in garden what blind foolish man has done in a few years, will deny an all-seeing being in thousands of years could effect (if the Creator chose to do so), either by his own direct foresight or by intermediate means.—which will represent the creator of this universe. (DeBeer 1958, 45)

In this passage we see a very powerful expansion of the scattered images and runs of explanatory notes which characterized the notebooks. Undoubtedly, in the "sketch," Darwin is not merely presenting research but is engaged in the process of persuasive exposition. Most impressive is the way he has developed a compound image—of a demiurge," a being infinitely more sagacious than man (not an omniscient creator)"—to superintend the Malthusian process without directly controlling it. This image is neither God nor man, neither the breeder nor yet nature, but an explanatory fiction that incorporates parts of all of these thereby enabling Darwin to coordinate a complex argument that otherwise threatens to spin out of control. The demiurge allows Darwin to bring unstaunchable variation, Lyellian time, and geologic, geographic drift together with Malthusian population pressure. Whatever the exact conceptual theological meaning of this passage, the important point illustrated is that Darwin's figure, "natural selection," is not primarily an insight into technical biology, it is primarily an insight into a problem of popular communication.

To confirm this point, we can compare it to A. Wallace's formulation, whose insights into evolution also came from Malthus—but with no analogy to domestication. In applying Malthus to biology, Wallace observed, "The action of this principle is exactly like that of the centrifugal governor of the team engine" (ibid., 278). The problem with this image rhetorically is similar to the problem of Darwin's initial wedge image; it is unacceptably accurate. "The centrifugal governor on a steam engine" was a thermostat and this same notion of a self-acting system sustained in motion by pressures internal to the system is what also is conveyed by Darwin's image of nature as an interconnected series of wedges.

Not only does Wallace's image differ from Darwin's "natural selection," Wallace also begged Darwin to drop it! In the very letter in which Wallace brought to Darwin's attention H. Spencer's suggestion of the phrase "survival of the fittest," Wallace urged Darwin to abandon the figure "natural selection" on the grounds that the figure inevitably suggested an active agent doing the selecting (letter to Darwin, 2 July 1866, F. Darwin 1903, vol. 1, 267–70). Darwin's response to Wallace is illuminating. With certain modest demurrers, Darwin agrees with Wallace's basic critique, notes that it is really too late to change—this was in 1866—and ends by noting that a natural selection also operates in language. He then appeals to he arbitration of this larger cultural selection to justify his expression (letter to Wallace, 5 July 1866, ibid.).

Though much more might be said about Darwin and the rhetorical significance of natural selection, I will make but one further observation before moving on to the *vera causa* principle and the arrangement of Darwin's book. After the first edition of the *Origin,* the passage about wedges—which initially had been included—no longer appears. He most likely dropped it, not because he changed his mind and now thought it was inaccurate, but because—along with other illustrations he dropped or included for persuasive reasons—he thought it was drawing too much heat.[1] Darwin's extensive record of audience adaptation clearly suggests that "natural selection" was coined in the words of E. Manier, as "place-holding allusion" (1978, 174); it established a semantic-rhetorical halfway house between mechanism and miracle.

The Vera Causa Principle, Natural Theology and the Arrangement of Darwin's Chapters

In examining the development of Darwin's trope "natural selection," we have jumped ahead of our story. In the *Origin*, "Natural Selection" is the fourth chapter—a chapter Darwin referred to in a letter to his publisher Murray as "the key-stone of my arch" (F. Darwin 1911, vol. 1, 511). In examining the image "natural selection" first and in textual isolation, we have skipped over the careful stonework which leads up to and follows from it. Hodge has compared the three-part *vera causa* logic of Darwin's original plan for the *Origin* to a "concerto" (1977, 242). While I agree with him in theory, the more I read, the more the *Origin* seems organized like a fugue.[2] Within the first four chapters (even as early as the end of the first), we have in place the basic themes and strategies which will characterize the rest of the book. From then on, indeed no later than the last section of the introduction to chapter four, we have a series of topical variations on a theme which are used to organize specific subject domains. In short, after the introduction to chapter four, I do not see the arrangement of the *Origin* going anywhere.[3] Instead of a progression, each chapter rehearses its own version of "existence," "competence," and "responsibility." With certain exceptions, one can almost read any chapter in isolation and gain a fair sense of Darwin's argument as a whole.

To offer at least a partial test for this interpretation, let us briefly examine key thematic elements of the first chapter, which falls easily into three parts. A first untitled run of seven pages addresses the fact of variation under domestication, establishes the interrelation between domestication and nature, and introduces the notion of "unconscious selection." A middle section of eight-and-a-half pages is "On The Breeds of the Domestic Pigeon," and a final thirteen-and-a-quarter-page section examines "Selection" as practiced consciously by English breeders and "unconscious selection" as practiced by "savages."

Here are some interesting statistics. As one might expect, the first chapter shows a high incidence of the term "domestication." The word appears 74 times in the chapter's 36 pages. What one might not expect is the high incidence of the term "nature," which appears 24 times. When we add to this the 36 references to "wild," the resulting 60 references to the natural world roughly indicates how thoroughly "domestication" and "nature" are interconnected in this chapter.

Chapter one is the thinnest edge of a strategic-tactical wedge. If the reader can be persuaded here that domestic species manifest virtually unlimited variability, were descended from one or a very few aboriginal wild breeds, breed true, and yet vary among themselves more even than species in nature, then that reader will not be well prepared to resist the a fortiori logic of the topic "greater and less"—whatever man can do nature can do better—which Darwin will make increasingly explicit in subsequent chapters. By chapter two the reader begins to realize the implications of choices already made, and implicitly rehearsed in chapter one.

Two central features of Darwin's "attack" on the argument from design that chapter one sets in place need to be noted from the start. First, the question of "creation" is subtly but decisively redefined. There is no question of an absolute beginning, either of life itself or of any particular structure, instinct or adaptation. The first "existence" question thus that Darwin establishes is the "existence" of a continuum of variation. This is evident in the first two sentences of the chapter:

When we look to the individuals of the same variety or subvariety of our older cultivated plants and animals, one of the first points which strikes us, is, that they generally differ much more from each other, than do the individuals of any one species or variety in a state of nature. When we reflect on the vast diversity of the plants and animals which have been cultivated, and which have varied during all ages under the most different climates and treatment, I think we are driven to conclude that this greater variability is simply due to our domestic productions having been raised under conditions of life not so uniform as, and somewhat different from, those to which the partent-species have been exposed. (Darwin 1967, 7)

A continuum of variation is established, as really existing, or as possible or probable; a series of intermediate steps are shown or postulated to account for how one might start from a simple beginning and arrive at a complex end. Darwin will use the same technique to account for the geometric structure of the combs of bees, the structure of the eye, the sterility of certain insects, or the development of the habit of ants making slaves.

In this particular passage, nature is mentioned in the first sentence and, by the end of the second sentence, domestication has been folded into it. The only difference "we are driven to conclude" between the unstaunchable variation in domestication and the relatively less patent variation in nature is due to a "simple" cause—"domestic productions" have been raised under conditions different from those of their parent species.

To underscore the competence of heredity to pass on variation Darwin affirms a maxim at once Biblical and agrarian and enfolds a central piece of folk wisdom under the banner of change, "No breeder doubts how strong is the tendency to inheritance: like produces like is his fundamental belief: doubts have been cast on this principle by theoretical writers alone" (ibid., 12). From this and similar passages Darwin dissociates himself from "theoretic" writers who cast doubt on experience and associates himself with practitioners. In the process of associating with practice and convention, Darwin, in the spirit of the incremental variants he is discussing, freights "common sense" with theory. Appealing to the reader's store of anecdotal knowledge, Darwin observes, almost in the spirit of gossip, "Everyone must have heard of cases of albinism, prickly skin, hairy bodies, & c.; appearing in members of the same family. If strange and rare deviations of structure are truly inherited, less strange and common deviations may be freely admitted to be inheritable" (ibid.). Clinching his "from greater to less" appeal, Darwin concludes, "Perhaps the correct way of viewing the whole subject, would be, to look at the inheritance of every character whatever as the rule, and non-inheritance as the anomaly (ibid.).

By the end of the first section Darwin has established the existence of an unstaunchable flow of variation in domestication, and hinted at the same in nature. Most importantly he has prepared the way to present a central feature of this theory—the mechanism of organic change "selection," not as a theory, but every bit as much as Paley's watch and "contrivances" as a datum of practical experience and common sense.

The short middle section of the chapter "On the Breeds of the Domestic Pigeon" gives Darwin an opportunity to establish what Perelman would call "presence" for his theory. Here Darwin creates the illusion—from data too near home to deny—that the reader (at least in domestication) has seen evolution in action. In this example the reader, already supplied with evidence of the existence of variation in domestication and in nature, will now see not only that variation exists, but "selection" is both "competent" and "responsible" for the production of the nineteen different domestic breeds of pigeon from a single wild form—the *Columba livia* or common rock pigeon.

Darwin uses the pigeon example to redefine the meaning of "design" in nature and to solidify the relation of this new meaning to seemingly prereflective "commonsense" experience. At the end of his exposition, Darwin proceeds to make himself the mediator of an epistemological showdown between theory insufficiently attuned to practical experience and practical experience insufficiently observant about its own theoretical significance. The pivotal point of his detailed example is the following:

> One circumstance has struck me much; that all the breeders of the various domestic animals and the cultivators of plants, with whom I have ever conversed . . . are firmly convinced that the several breeds to which each has attended, are descended from so many aboriginally distinct species. . . . The explanation, I think is simple: from long-continued study they are strongly impressed with the differences between the several races; and though they well know that each race varies slightly, for they win their prizes by selecting such slight differences, yet they ignore all general arguments, and refuse to sum up in their minds slight differences accumulated during many successive generations. (Ibid., 28, 29)

Then he proceeds to spell out its evolutionary implications:

> May not those naturalists who, knowing far less of the laws of inheritance than the breeder, and knowing no more than he does of the intermediate links in the long line of descent yet admit that many of our domestic races have descended from the same parents—may not they learn a lesson of caution when they deride the idea of species in a state of nature being the lineal descendants of other species? (Ibid., 29)

The symmetry between the mistaken theory of the breeders and the ignorance of practice of the theorists is perfect. Each faithfully mirrors what K. Burke calls the "trained incapacity" of the other. And again Darwin emerges, as he has in earlier passages, as the mediator between extreme positions—with evolution representing the position of moderation and caution!

In the final section of the chapter Darwin once again sides with practice and through his questions leads readers to apply what they have learned of the breeder analogy to the central issue of the chapter and, indeed, of the book. Noting the vast differences among the breeds of the dog, of game cocks, and the multitude of various forms from the garden, Darwin affirms that "[w]e cannot suppose that all of the breeds were suddenly produced as perfect and as useful as we now see them; indeed, in several cases, we know that this has not been their history" (ibid., 30). Against the backdrop of a slightly varied example Darwin repeats the lesson dramatized earlier in his epistemological contrast between the breeders and the naturalists, "The key is man's power of accumulative selection: nature gives successive variations; man adds them up in certain directions useful to him. In this sense he can be said to make for himself useful breeds" (ibid.). Here, at the beginning of his discussion of "Selection," the existence, competence, and responsibility of variation as guided by selection to produce domestic breeds has been established (at least synecdochically) and, equally importantly, has been so in a way that affirms the priority of practical experience as the ground of theory. In effect, Darwin has out-Paleyed Paley by making of the power of the domestic breeder, as manifest in the derivation of the domestic pigeons, his counterversion of Paley's watch.[4] The example of the descent of the various pigeons from *Columbia livia* has become the paradigm of how things organic are "made." Since, as the first two sentences of the chapter have made evident, no essential difference lies between variation under domestication and under nature, essential features of the argumentation of the remainder of the book are already in place.

One final point needs to be made of the third section of chapter one. Even as the chapter began by establishing a continuum of variation, it ends by linking the continuum of variation to a continuum of consciousness. Selection, it seems, not only deals with variation, but varies itself. In civilized humanity selection is a fine art:

Not one man in a thousand has accuracy of eye and judgment sufficient to become an eminent breeder. If gifted with these qualities, and he studies his subject for years . . . he will succeed . . . if he wants any of these qualities he will fail. Few would readily believe in the natural capacity and years of practice requisite to become even a skilful pigeon-fancier. (Ibid., 32)

But even among "savages" Darwin observes that:

any one animal particularly useful to them, for any special purpose, would be carefully preserved during famines and other accidents, to which savages are so liable, and such choice animals would thus generally leave more offspring

than the inferior ones; so that in this case there would be a kind of unconscious selection going on. (Ibid., 36)

With his continuum of variation, Darwin has positioned himself to explain complexity by locating it on a relative scale rather than against
the mysterious backdrop of the unknown absolute divine beginning.
By his appeal to common sense, Darwin has captured for his own use
one of the central features of his opponent's traditional appeal. By his
continuum of consciousness Darwin has positioned himself either to
associate selection with intention or let it sink into the unconscious-
ness of nature itself.

Conclusion

While many conclusions might be drawn from what we have examined, here are three. First, what is most remarkable about Darwin's
argumentation as manifest in both his personification of the breeder
and in his localization of the *vera causa* principle is his commitment
to the concrete. In numerous instances—in his account of the complex
interrelations of being to being in his account of the bees, the cats, and
the heartsease, and in his many accounts of complex organs, instincts,
and relations—we gain a glimpse of how the rhetoric of science—in
moments of its very perfection—brings us to the end, in the sense of
the extreme limit, of the rhetoric of science.

If the critical point in Darwin's conversation with Hensleigh and
Elizabeth, recorded in his 1839 note, was his inability to make them
see the importance of small variations or to grasp the evolutionary
logic of Malthus, much of his thought from that point forward was
directed toward presenting an impressive abstract argument as though
it were a concrete datum of everyday experience. In Darwin's hyperde-
velopment of the dubious analogy between domestic breeding and
Malthusian population dynamics we see an aspect of the human
knowledge project that is in turn, or all at once, comic, rhetorical, and
tragic. It is comic or tragic for thought that must be expressed in concrete images appropriate to a particular audience at a particular place
and time inevitably will be a long study in what Whitehead (1964)
called "the fallacy of misplaced concreteness" (pp. 52, 52–56). To call
Darwin's insight the theory of "natural selection" is already to misrepresent it, but that cannot be helped for that is his representation. To
reflect on how to put it better is to return to Darwin's original prob-

lem. In the midst of the tragicomedy of knowledge a philosophic rhetoric might be understood as aimed at recovering the margin of cultural gain. Perhaps every historically significant advance, whatever the appropriate measure, requires or rests upon some great exaggeration.[5] Darwin's exaggerations are worth understanding—and exposing. Individual readers attuned to the rhetorical dimension of experience might see through the inevitable condescension or inequality between reader and author that such great exaggerations require. On the other side of deception, philosophic friendship may be achieved with the author. As Eliot more elegantly put it "[We] are only undeceived of that which deceiving can no longer harm" (1973, 26). Certainly the victory of Darwin's book is a victory not of science only but of scientific naturalism, a philosophy that requires a priori all taxonomic gaps be filled and all arguments for design be inadmissible.

Second, if rhetoric on the one hand necessarily distorts so that its message may reach a particular historical audience, on the other hand it also invites us to expand our notion of logic, or to recognize the ways in which well-established canons of formal reasoning may become rhetorical tactics. As Hodge has clearly shown, Darwin conceived of the organization of his work as an unfolding of the *vera causa* logic he had learned from Herschel and Lyell. What I hope my own analysis has added, partial though it necessarily has been, is that Darwin's departure from the *vera causa* logic is not simply an imperfection or an unnecessary muddying of a pattern that would have made his case clearer, it is also the consequence of his pursuit of a supplementary or alternative rhetorical logic. To the extent that audience adaptation becomes a dominant motive in a work, and particularly in Darwin's case where the conventions of natural theology were common both to his more popular and more technical audiences, his expository needs no longer could be met adequately by the more leisured logic of the concertolike structure of the *vera causa* principle. By tactically bringing forward the competence and responsibility steps and by topically melding them into the developmental fabric of his case, Darwin sought to meet other and more urgent audience expectations from traditional natural theology. The cumulative effect of Darwin's persistent repetition of small tactical moves set in place in the first chapter of the book and elsewhere eventually shifted the strategic center of gravity of the book from the arrangement of the chapters as a whole to the arrangement of each chapter individually.

Finally, a central contribution of rhetorical criticism to the history and philosophy of science is its technical contribution to understanding the complex logic(s) of historical change. Particularly important here is its identification of the communication conventions, for example, the constraints of prior genres and audience expectations, which are or may be turned against their original function and enlisted as instruments of change. One of the strongest potential contributions of a rhetoric of science will be in showing how significant scientific texts, the logical structure of which are well understood, may yet yield fresh insight when read as performances and when the principal figures in scientific revolutions are understood as performing artists. The principle of performance—of complex adaptation to complex and refractory human situations—should lead us not to expect perfection or complete consistency in the logic or the rhetorical art even of singularly great texts such as the *Origin*. Rather our focus should be on how well particular individuals brought the resources of significant insight to bear on the varied fabric of tradition to effect significant change. No more with rhetoric than with logic can we equate success with simple persuasion; our focus must be on invention—discovery of the possible. Few things in history are less stable than the principle of sufficient reason and the language which embodies it is rarely given but hermeneutically and rhetorically must be reconstituted by art. Examining that process in concrete situated cases is a central task, and potential contribution, of what I take to be the rhetoric of science.

NOTES

Tables 1, 2, and 3 in figures 9.1–9.3 are reproduced with the permission of the Council of the British Society for the History of Science. They were first published in *The British Journal of the History of Science* 22 (1990): 66–90.

1. See Darwin's letter to Harvey, August 1860, "The bear case has been well laughed at and disingenuously distorted by some into my saying that a bear could be converted into a whale. As it offended persons, I struck it out in the second edition; but I still maintain that there is no especial difficulty in a bear's mouth being enlarged to any degree useful to its changing habit . . ." (F. Darwin 1903, vol. 1, 162–63). Darwin also included points (in addition to his allusions to Gray's theology) because others believed in them—even though he firmly rejected them. See his letter to Lyell, 1 September 1860, "I put in the possibility of the Galapagos having been continuously joined to America, out of mere subservience to the many

who believe in Forbes's doctrine, and did not see the danger of admission of small mammals surviving there in such a case. The case of the Galapagos . . . in fact convinced me more than in any other case of other islands, that the Galapagos had never been continuously united with the mainland; it was mere base subservience, and terror of Hooker and Co." (F. Darwin 1911, vol. 2, 127).

2. By saying the structure of the *Origin* is "fugueal," I mean that strategically, at least from the standpoint of a putative "general reader," I find the *Origin* roughly organized by a series of "short subjects or themes." While I cannot exactly say these are "harmonized according to the laws of counterpoint and introduced from time to time with various contrapuntal devices" (*Oxford English Dictionary*, 585), I can say that these themes are repeated in a variety of subject domains—and do not lead to a distinct thematic climax. Thus, in my reading, while there is a progression of subjects in the *Origin*, they are treated by a simultaneous repetition of strategies.

3. However, as J. Lennox points out, a specialist reader could appreciate these—and for this reader the *Origin* would go somewhere. From the standpoint of a general reader, Darwin's new departures provide fresh fields in which to rehearse basic tactics of argument adumbrated in the first chapter, or at most by the first four. It is very difficult for a general reader to sense a thematic progression beyond the first four chapters of the *Origin*—hence my "fugueal" counter to Hodge's "concerto."

I do see one rhetorical development in Hodge's "competence" chapters 4–8 (particularly in chaps. 6–8) that are tied to a crucial issue which he faces in his "responsibility" chapters 9–13. Chapters 6–8, following the model of the breeder in chapter one, offer a plethora of examples in which readers are urged to extrapolate small variations across sometimes small and sometimes very large gaps in the evidence. An uncritical reader might be convinced that natural selection was invincible; a skeptical one might recognize Darwin had at least battled convention to a draw. These chapters prepare the reader for chapter nine, "On the Imperfection of the Geological Record," in which Darwin has to convince the reader that the empirical evidence, not his theory, is imperfect. Once Darwin's pattern of extrapolating small effects over small and large gaps has become second nature to the reader, Darwin's reader may gradually come to recognize that the theory must in principle be true and all objections, no matter how great, are products of "imagination" and "cannot be considered real" (Darwin 1967, 459). This is a significant development but it is fully adumbrated—though certainly not all the persuasive effects Darwin will garner from it are yet evident—in Darwin's face-off between the breeder and the naturalist in chapter one.

Furthermore, I do not entirely take Darwin's statement toward the end of chapter five, after his chart of the hypothetical operation of natural selection, at face value. Darwin claims, "Whether natural selection has really thus acted . . . must be judged of by the general tenour and balance of evidence given in the following chapters" (ibid., 127). But within the next few lines, this warning seems purely formal and indeed empty, "But we already see how it entails extinction; and how largely extinction has acted in the world's history, geology plainly declares. Natural selection also leads to divergence of character. . . . Thus the small differences distinguishing varieties of the same species will steadily tend to increase till they come to equal the greater differences between species of the same genus, or even of distinct genera" (ibid.). Hypothetical and schematically idealized though Darwin's diagram is, he undoubtedly regards the relations it depicts as real. As this statement clarifies, the metaphysical premise of a radical naturalism implicit in his images and examples leads the reader to the consideration of the further evidence with something less than an open mind. When the reader is told at the end of chapter five that refusal to imagine a common ancestor to the horse, the hemionus, the quagga, and zebra is equivalent to believing with the ancient cosmogonists that fossil forms found in mountains were specially created to mock the living forms on the seashore and that denial of descent "makes the works of God a mockery and a deception" (ibid., 167) Darwin merely emphasizes a theme already in place before the end of chapter one. Later chapters develop these strategies in a variety of important contexts. For a fuller account, see Campbell (1995).

4. That Darwin is almost literally taking a page from Paley's book is indicated by the following passage. Having insisted earlier on the necessary relation between "contrivance" and "contriver" (Paley 1863, 10–12) and that not even a watch that would make other watches would change the force of this argument, Paley extends his argument to physical laws in general. One passage in particular is reminiscent of the earlier passage in Darwin's 1842 manuscript—and echoes again strongly in Darwin's account of how breeders make domestic breeds, "because the Deity, acting himself by general laws, will have the same consequences upon our reasoning, as if he had prescribed these laws to another" (ibid., 27). (I thank P. Nelson for drawing this passage to my attention.)

5. Darwin's exaggerations were indeed great. Not only did he claim that were it to do over again he would drop the term "selection" in favor of "natural preservation" or "naturally preserved" and reply to Wallace's objections that usage would be decided by the natural selection that operates in language, but he added theological language to later editions of the Origin, and financially underwrote and helped distribute Gray's pamphlet, "Natural Selection Not Incompatible With Natural Theology," (in which he emphatically disbelieved). The theory of "special creation" which plays so large a part in the Origin is also largely his invention. As D. Hull (1983) points out, "Critics both at the time [1859] and since agree with Rudwick . . . that in arguing against the special creationists, Darwin 'presented his theory with only a straw man to oppose it: either slow trans-specific evolution by means of "natural selection", or direct divine creation of new species from the inorganic dust of the Earth' " (p. 63).

REFERENCES

Barrett, P.; P. Gautrey; S. Herbert; D. Kohn; and S. Smith, eds. 1987. *Charles Darwin's Notebooks, 1836–1844*. Ithaca: Cornell University Press.

Bowen, F. 1860. "Darwin on *The Origin of Species.*" *North American Review* 90: 474–506.

Campbell, J. A. 1995. "The Comic Frame and the Rhetoric of Science: Epistemology and Ethics in Darwin's *Origin.*" *Rhetoric Society Quarterly.* In press.

Darwin, C. 1967. *On the Origin of Species: A Facsimile of the First Edition.* New York: Atheneum.

Darwin, F., ed. 1903. *More Letters of Charles Darwin,* vols. 1–2. London: John Murray.

———. 1911. *The Life and Letters of Charles Darwin,* vols. 1–2. New York: Appleton.

DeBeer, G., ed. 1958. *Charles Darwin and Alfred Russel Wallace, Evolution by Natural Selection.* Cambridge, England: Cambridge University Press.

———. 1963. *Evolution by Natural Selection.* Cambridge, England: Cambridge University Press.

Eliot, T. S. 1973. *The Four Quartets.* New York: Harcourt, Brace & World.

Gould, S. J. 1977. *Ever Since Darwin.* New York: Norton.

———. 1980. *The Panda's Thumb.* New York: Norton.

———. 1983. *Hens' Teeth and Horses' Toes.* New York: Norton.

Gray, A. 1962. *Darwiniana.* Edited by A. Hunter Dupree. Cambridge, Mass.: Harvard University Press.

Hodge, M. J. S. 1977. "The Structure and Strategy of Darwin's Long Argument." *British Journal of the History of Science* 10: 237–46.

Hull, D. 1983. "Darwin and the Nature of Science." In D. S. Bendall, ed., *Evolution from Molecules to Man.* Cambridge, England: Cambridge University Press, pp. 63–80.

Kavaloski, V. C. 1974. *The Vera Causa Principle: A Historico-Philosophical Study of a Metatheoretical Concept from Newton through Darwin.* Ph.D. dissertation, University of Chicago.

Kohn, D. 1980. "Theories to Work By: Rejected Theories, Reproduction and Darwin's Path to Natural Selection." *Studies in the History of Biology* 5: 67–170.

Kohn, D., and M. J. S. Hodge. 1985. "The Immediate Origins of Natural Selection." In D. Kohn, ed., *The Darwinian Heritage.* Princeton: Princeton University Press, pp. 185–206.

Limoges, C. 1970. *La selection naturelle: Etude sur le premier constitution d'un concept.* Paris: Presses Universitaires De France.

Manier, E. 1978. *The Young Darwin and His Cultural Circle.* Dordrecht: Reidel.

Paley, W. 1863. *Natural Theology.* Boston: Gould & Lincoln.

Stauffer, R. C. 1975. *Charles Darwin's "Natural Selection," Being the Second Part of His Big Species Book Written from 1856 to 1858.* New York: Cambridge University Press.

Whitehead, A. 1964. *Science and the Modern World.* New York: Mentor Books.

Comment: Darwin's Recapitulation

James G. Lennox
Department of History and Philosophy of Science, University of Pittsburgh

J. Campbell rightly judges *On the Origin of Species* (Darwin [1859]
1964) a rhetorical tour de force, wisely sculpted from a much larger
and less attractive precursor, to persuade. To understand its persuasive
power, however, one must identify Darwin's intended audience. Three
questions in particular seem crucial to such identification. *Who* was
he trying to persuade, *of what* was he trying to persuade them, and
what would be persuasive *to them?* Substantial overlap occurs between Campbell's answers and my own. I recognize that Darwin had
primary and periferal audiences in mind, the primary consisting of the
influential members of a few leading scientific societies in Great Britain. He wished to persuade them that he had a plausible theory about
how species originate that appealed *to secondary causes alone.* This,
he knew, was the holy grail of his audience. The *Origin*'s introduction
(ibid., 1) refers immediately to a "great philosopher" who calls the
species question the "mystery of mysteries"—this is J. Herschel, the
phrase coming from a letter to C. Lyell in which he expresses his belief
that species must originate by secondary, not miraculous, causes, a
point on which Lyell emphatically agrees (see Barrett et al. 1987, 59;
Cannon 1961).

What sort of a theory of naturalistic species origination would be
persuasive to Darwin's primary audience? Here Campbell and I agree
that Darwin needs to establish "the principle of selection" as a *vera
causa* of species origins if he wishes to succeed with this audience.

Sharing this framework with Campbell, I nevertheless find much

237

with which to disagree in his characterization of how Darwin carries out this rhetorical enterprise, and the origins of its principle literary devices. Reversing Campbell's order, I begin with his claim that "after the introduction to chapter four, I do not see the arrangement of the *Origin* going anywhere. Instead of a progression, each chapter rehearses its own version of 'existence,' 'competence,' and 'responsibility.' "

To say that Darwin mounts a *vera causa* strategy in the *Origin* is to say very little—specifically, it is to say that he appeals to causes known to exist, competent to explain the phenomena. The real issue of interest is how one supports such claims of existence and competence. In the first four chapters of the *Origin,* Darwin's strategy is to identify a familiar process, competent to produce well-marked varieties from a common stock. He finds such a process in the selective practices of plant and animal breeders. He seeks to extend this principle beyond its established field of competence to the production of species. How does he describe his own move? "How will the struggle for existence, discussed too briefly in the last chapter, act in regard to variation? Can the principle of selection, which we see is so potent in the hands of man, apply in nature? I think we shall see that it can act most effectually" (Darwin [1859] 1964, 60).

This had better not, however, be a merely *metaphoric* extension— if so, his audience will not accept it. Darwin is not, contra common assumption, building the metaphor of natural selection by analogical reasoning; he generalizes and extends the principle of selection from a domestic to a natural context. There is, he says, a single principle: *selection.* It has been shown to be competent to produce varieties in human hands. Furthermore, the *Origin*'s first two chapters have shown that trustworthy naturalists cannot distinguish well-marked varieties from insipient species, which means that in human hands we might as well say it has been shown to have the power of producing species. Now the question is whether *selection* can so act in nature. Darwin, moreover, by cleverly juxtaposing two different questions as if they were equivalent, makes it clear that this principle applies to nature *by virtue of* the way the struggle for existence acts in regard to variation. So the struggle for existence, which chapter three argues is a corollary of Malthus's principle of population, accomplishes what human artifice accomplishes in the breeding pens. Note how he first introduces the principle in detail, early in chapter three, "I have called this principle, by which each slight variation, if useful is preserved, by the term of Natural Selection, in order to mark its relation to man's power of selection" (ibid., 61).

Again, no mention of metaphor is made (as there is, explicitly, when he discusses the struggle for existence). Rather, there is one principle—the preservation of useful variations. It is known to be potent in one domain; Darwin seeks to persuade his readers that, because of the struggle for existence in nature, that principle applies to species in nature as well, and is similarly potent.

This is one acknowledged way of establishing a *vera causa*—the extension of a known, competent principle to a new domain. But unlike Campbell, I find evidence that Darwin sees this as a mere first step. In the summary of chapter four, Darwin admits that "whether natural selection has really thus acted in nature . . . must be judged by the tenour and balance of evidence given in the following chapters" (ibid., 127). That is, the evidence for the existence of *natural* selection is yet to come—read on! On Campbell's reading, according to which the last chapters merely rehearse the strategy of chapters one to four, this claim makes little sense.

In fact, I insist on *two* independent strategies adopted from this point on, aimed at different audiences. In chapters 9–13, Darwin goes for explanatory unification (as some of us say) or consilience, as W. Whewell would say. In correspondence Darwin constantly insisted—comparing natural selection to the standard examples of the wave theory of light and gravitational attraction—that the true test of whether one had a *vera causa* is in a theory's ability to "group and explain large classes of facts" (1972, 9; see also letters to A. Gray, F. W. Hutton, and C. Lyell cited by D. Hull 1973, 44–45). For Whewell, this was the most compelling evidence for a *vera causa*, but it was equally dominant in the Herschellian account of confirmation (see Ruse 1975). With P. Kitcher (1985) and others, I see in chapters 9–13 a related, but very distinct strategy: to show that already agreed-to facts in geology, biogeography, classification, instinct, and so on—facts which no one connected with species origination—to show that descent under selection could explain them better than any competitor.

Chapters 6–8, on the other hand, adopt a third strategy, one which Darwin apparently learned directly from reading Lyell's *Principles of Geology* (1832): In the face of arguments that one's causal theory *could not possibly* explain X, tell an imaginative narrative—a "Just-

So Story," if you like—with your causal mechanism as the chief protagonist and X as a perfectly natural result of its iterative operation. I
call them "Darwinian Thought Experiments," and R. Brandon calls
them "how-possibly explanations" (Lennox 1991; Brandon 1990,
176–83). They are aimed to disarm the antagonistic critic, but also to
reassure the timid supporter that the antagonistic critic can be dis-
armed.

In short, then, I see what happens after chapter 4 as the mobilization
of two distinctive and complex argumentative strategies. And in each
subdomain, of course, are substrategies. A book could be written on
how appealing the marshalling of the geological and the biogeographi-
cal facts would have been to, say, Lyell.

What of the "trajectory of discovery"—if natural selection be imagined as the endpoint, how did Darwin get there? Here Campbell turns
to the *Transmutation Notebooks* of the 1837–1838 period (see Barrett et al. 1987). Again I am in substantial disagreement with Campbell. Rather than see Darwin's way to natural selection leading *from*
Malthus *to* the Breeder, I see the reverse. And by so seeing it, I provide
a different account of the reader-friendly character of the *Origin*—it
is a recapitulation, shared with the reader, of Darwin's own path of
discovery.

In Campbell's reconstruction, Darwin tries various metaphors—
most infamously "Nature as a block full of wedges"—before gradually hitting on domestic breeding as a model for selection in nature. I
find this an unlikely scenario since the issue of whether breeders could
create new species was central to the pre-Darwinian debate over species transmutation, as evidenced in Lamarck and in Lyell (1832, 26–
35), the texts that introduced Darwin to the issue. It is thus anteced-
ently likely that Darwin would have had the breeder's ability to create
distinct varieties out of common stock as a starting point of his inquiry.

And indeed, this is what the *Species Notebooks* reveal. Darwin begins discussing his reading of the pamphlets and manuals of domestic
breeders in Notebook C, well before the famous Malthus entry late in
Notebook D:

Mr. Yarrel.—says my view of varieties is exactly what I state.—&c picking
varieties. Sir J. Sebright first got point on hackles on Bantams by crossing with
common Polish cock . . . & then recrossing off spring. till size diminished,
but feathers continued by picking chickens of each brood.—These bantam

feathers at last got ducky, then took white Chinese Bantam crossed & got some yellow & others yellowish & white varieties by picking the yellow one & crossing with duck bantams procured old variety. (Barrett et al. 1987, 275)

Both of these examples were later used in Darwin (1972)—Sebright's breeding pamphlets being favorites of Darwin's on the topic. Here Darwin is well aware that the new varieties are created by picking the animals with the desired trait at each generation and crossing them.

In a later passage, Darwin returns to Sebright's work:

Sir J. Sebright—pamphlet—most important, showing effects of peculiarities being long in blood—thinks difficulty in crossing race.—bad effects of incestuous intercourse.—excellent observations of sickly offspring being cut off—so that not propagated by nature.—Whole art of making varieties may be inferred from fact stated.—(Barrett et al. 1987, 279)

Campbell notes, in much later entries in Notebook E, Darwin's use of "the breeder's term 'picking' for the more elegant 'selection.' " Alas, it was apparently Darwin who lacked the elegance of the breeders. Let us note a passage, scored in Darwin's copy, in which Sebright uses the term "selection" precisely in the manner later adopted by Darwin, "If a breed cannot be improved, or even continued in the degree of perfection at which it has already arrived, but by breeding from individuals, so selected as to correct each other's defects, and by judicious combination of their different properties" (quoted in Barrett et al. 1987, 279). Sebright continues, "A severe winter, or a scarcity of food, by destroying the weak and the unhealthy, has had all the good effects of the most skillful selection" (ibid.).

So, then, Darwin studied and speculated on domestic selection at this point well before he realized the relevance of Malthus's principle of population to his theory—though surely comments like the last quoted from Sebright helped Darwin *see* the relevance of Malthus! He is at this point in Notebook C referring to his work as "my theory". Thus, the evidence favors that the "way to natural selection" runs *from the Breeder to Malthus,* not vice versa.

What about the enigmatic "wedge" metaphor and its equally enigmatic disappearance? Campbell claims it is the "biological counterpart to the mindless, directionless mechanisms of uniformitarian geology." This is important to his portrayal of the rise of natural selection, which he sees as a *metaphorical subversion* of design. I have my doubts. In its first appearance, in the *Notebooks*, the wedges are "*try-ing* to force every kind of adapted structure into the gaps in the econ-omy of Nature." Likewise in every subsequent appearance—the 1842 "sketch," the 1844 "essay," the first edition of the *Origin*—the wedges are "being forced," "driven inwards by incessant blows." All meta-phoric, no doubt. But the reference to a conscious agent striking the blows is always present, as much as it is in the idea of the struggle for existence "selecting and scrutinizing" slight differences.

As Campbell notes, this metaphor is dropped after the first edition, having been with Darwin for 20 years. Why? Campbell would argue that it is less effective on Darwin's audience than the metaphor of se-lection with its apparent reference to conscious design. But for the rea-sons I mentioned, this is implausible. Furthermore, it is not a competi-tor of selection in any sense. This metaphor aids us in grasping the power of relaxing or intensifying a Malthusian "check" on popula-tions. Natural selection, on the other hand, refers to a *consequence* of the struggle for existence generated by Malthusian population dynam-ics. I find this metaphor unhelpful because of its image of the natural world as a homogenous block of wood. Darwin may well have come to this viewpoint as well.

Finally, the differences between Darwin and Wallace's theories seem to work against Campbell's reconstruction of Darwin's path of discovery. The idea of selective crossbreeding as a means of producing well-marked varieties is, as we have seen, familiar to Darwin from his early years. It is a trope that never impacts Wallace. The Malthusian insight could be directly applied to the problem of adaptation. For Darwin, however, the Malthusian insight caused the problem of how it fits with the breeder-inspired principle of selection.

Wallace, along with many others, tried to convince Darwin that, in fact, it did *not* fit. But it was dear Charlie's baby, and he protected it from friend and foe.

On this reading, the trajectory of *Origin*, chapters 1–4—from varie-ties produced by domestic selection, to variation in nature and the tax-onomic anarchy of the experts who describe it, to Malthus and the Struggle for Existence, and finally to Natural Selection—constitutes a literary recapitulation of Darwin's own theoretical ontogeny—from the Breeding pamphlets, to *On the Principle of Population*, to Natural Selection. As Campbell says, we are in "the company of an author who welcomes the reader as his companion on the second voyage of intellectual discovery"—or perhaps the third, the second having been recorded in the *Transmutation Notebooks.*

REFERENCES

Barrett, P. H.; P. Gautrey; S. Herbert; D. Kohn; S. Smith, eds. 1987. *Charles Darwin's Notebooks, 1836–1844.* Ithaca: Cornell University Press.

Brandon, R. N. 1990. *Adaptation and Environment.* Princeton: Princeton University Press.

Cannon, W. F. 1961. "The Impact of Uniformitarianism: Two Letters from John Herschel to Charles Lyell 1836–37." *Proceedings of the American Philosophical Society 105:* 301–14.

Darwin, C. [1859] 1964. *On the Origin of Species: A Facsimile of the First Edition.* Cambridge, Mass.: Harvard University Press.

———. 1972. *The Variation of Animals and Plants under Domestication: The Works of Charles Darwin,* vol. 7. New York: AMS Press.

Hull, D. 1973. *Darwin and His Critics: The Reception of Darwin's Theory of Evolution by the Scientific Community.* Chicago: University of Chicago Press.

Kitcher, P. 1985. "Darwin's Achievement." In N. Rescher, ed., *Reason and Rationality in Natural Science: A Group of Essays.* Lanham: University Press of America, pp. 127–89.

Lennox, J. G. 1991. "Darwinian Thought Experiments: A Function for Just-so Stories." In T. Horowitz and G. J. Massey, eds., *Thought Experiments in Science and Philosophy.* Savage: Rowan & Littlefield, pp. 223–41.

Lyell, C. 1832. *Principles of Geology,* vol. 2. London: John Murray.

Ruse, M. 1975. "Darwin's Debt to Philosophy: An Examination of the Influence of the Philosophical Ideas of John F. W. Herschel and William Whewell on the Development of Charles Darwin's Theory of Evolution." *Studies in History and Philosophy of Science 6:* 156–81.

10

Rhetoric in the Context of Scientific Rationality

John Lyne

Department of Communication Studies, University of Iowa

This essay explores the possible role and function of rhetoric in the context of scientific rationality, assuming that some of the standard issues concerning scientific rationality itself can be bracketed or otherwise held at bay. I begin by accepting the well-advertised dangers of rhetoric, especially its capacity to woo with words and abrogate better reflective judgement. Plato explored the problem masterfully. Simply put, rhetoricians might generate the *appearance* of knowing what they are talking about and manage to persuade others, a skill to be avoided in the context of science. They may also play upon one's emotions or personal vulnerabilities. Again, such practices would be a danger to science. As any student of the rhetorical tradition knows, however, Plato made room for a noble rhetoric that did not trade on ignorance, but rather added impulse to truth. The two dialogues in which he writes about rhetoric, *Gorgias* and *Phaedrus,* culminate in vivid mythic images, driving home the point imaginatively, as if to show that literal truth is not enough. Needless to say, the dialogues are, in contour and strategy, rhetorically brilliant.

Much has changed since Plato's day, of course, especially in the realm of science. We now have a great deal more space for examining scientific discourses as situated within the scope of society and culture. I will venture some general claims concerning the role of rhetoric in the context of modern scientific rationality in the hope that they might serve as a prod to further discussion.

As Science Requires a Grammar and a Logic, So Too It Requires a Rhetoric

Linguists divide their turf into syntactics, semantics, and pragmatics. In noticing that accounts of philosophical rationality sometimes seem to divide according to the same three categories, I do not mean to suggest that science is just linguistics writ large. But I do find an interesting parallel. Rhetoricians recall the long dormant tradition of grammar, logic, and rhetoric—the ancient language arts trivium that was refashioned by C. S. Peirce under the rubric of Speculative Grammar, Critical Logic, and Speculative Rhetoric (Lyne 1980). This was reintroduced by C. Morris as semantics, syntactics, and pragmatics. The extent to which any of these triads might be considered to designate the same thing is, certainly, arguable. Peirce was not thinking about oratory when he proposed Speculative Rhetoric, for instance. Yet again, an interesting parallel remains. Rhetoric seems parallel to pragmatics, especially Peirce's Speculative Rhetoric, which would be a study of how habits of signification become adopted, embraced, and engrained in the habits of scientific investigators.

Some philosophers of science advocate a "semantic" model of science (Suppe 1989, Thompson 1989) in specific contrast to what they see as a dominant "syntactic" model in the philosophy of science. According to this conception, science does not depend on axioms and rules for their application, but rather on a family of models that interact with one another while purporting to hold a complicated relationship to the physical realities represented. Whereas a syntactic conception of theory is deductive and axiomatic, a semantic conception does not presume an integrated theory. Rather, it treats a theory as comprising a class of models, each with a domain presumed to be isomorphic to it. P. Thompson (1989) summarizes the main features of this approach:

(1) Theories are semantic structures which specify the behavior of systems, not the behavior of phenomena; (2) phenomena are not explained by deducing them from laws but by asserting that the phenomenal system of which they are a part is isomorphic to the physical system specified by the theory and hence has the same causal structure; (3) the existence and nature of an isomorphic relationship asserted to obtain between a physical system and a phenomenal system are not specified by the theory, and their establishment is

complex, requiring reference to theories of experimental design, goodness of fit, analysis and standardization of data, and so forth, and to causal chains based on other theories. (P. 82)

For Thompson, the theory of evolution consists of a family of inter-acting models, not a single self-consistent and unified deductive system. Biological scientists debate which of the models under consideration actually applies to the natural world, and how they apply. "Application" depends on various procedures and theories.

In respect to biological theories, at least, this semantic account seems to be more persuasive as an account of what happens in the development, testing, and extention of scientific theories than is the so-called syntactic model. The semantic approach recognizes the coexistence and interaction of a family of different theories, not just as a succession of one replacing another, but as a part of business as usual in the empire of biological science. For example, population genetics, biometric genetics, and molecular genetics operate upon different sets of assumptions, or, one might say, different models (Howe and Lyne 1992). To the extent that they operate at a distance from each other, the models need not compete. They need not even interact. I would highlight the problem of what description is to be given to the space in which such different models or theories do interact, as they sometimes do, especially in the context of grand synthetic theories which attempt to cast an explanatory net over all the biological sciences. As much as the semantic theory seems more satisfying than the syntactic theory, it does not seem to deal adequately with this question.

Here the analogy to linguistic activity can be instructive: Syntactics and semantics are in the service of pragmatics, even if it presupposes them. Attending to pragmatics should not be thought of as just the finishing touches on a complete syntactic and semantic account. Rather, it involves the situated management of meaning where one is trying to make a persuasive case. Such management has been studied historically under the rubric of rhetoric. Models and metaphors may be examined semantically, as Kuhn, Thompson, and others do. Insofar as one views metaphors and other language *pragmatically,* one has engaged their rhetorical dynamics, which involve both formal and nonformal elements.

While neglecting to give full credit to the complexity of the argument for the semantic theory, my understanding is that it remains concerned almost exclusively with questions of how models either fit with the world they purport to be isomorphic with or how they fit with each other. If a case is to be made for rhetoric in the dynamics of science, I believe that it can be made in large part in the space of interaction, where theories are considered by different investigators and presented to different audiences. This is the space, not just of testing and application, but of strategically managed meaning and reception. Enter rhetoric.

Rhetoric Applies Figuration and Implicit Argument as well as Literal Reference

Rhetoric is very much about the shaping of perspective. A model in science has some of the "taken from this perspective" quality that K. Burke and others have associated with metaphor. It is a "way of seeing" that can be more or less successfully applied, fecund, and so on, which is not to say it cannot frame empirically testable questions. Let me use this observation to pose the question of figuration, part of the rhetorical art. How do the multiple ways that we can position ourselves toward a subject matter influence our conception of what constitutes an adequate explanation of that subject matter? Rhetoric does different things with the same facts and theories because it plays upon them strategically, relative to audiences.

While other figures can be used, metaphor seems to have become a kind of synecdoche for a whole class of nonliteral devices—a kind of shorthand term for making the point that plain vanilla reference is only one of the many flavors we know how to produce. Being literal is only one of the games we play with language, and one that seems to make sense only as a kind of contrasting strategy, that is, to contrast with various nonliteral uses such as metaphor.

The cognitive value of a metaphor, according to many astute commentators, is in juxtaposing concepts from one familiar domain upon the content of another domain, thereby revealing, or suggesting, previously unnoticed properties. Metaphor is of considerable value to ordinary thinking and to intellectual progress in general because it is a means by which the familiar is used to map out the unfamiliar. In the humanities and social sciences its importance is widely recognized. In fact, "metaphor" is often the term used for a variety of framing notions found necessary even to hold cognition together or to get an inquiry going. Biological discourse is replete with metaphors too, of course, and not just to clarify meaning. To say that nature is red in tooth and claw is to invite an assessment of natural events as though they pivoted on predatory activity, and thus to counter metaphors of natural harmony or civility. Darwin recognized how easily his theory lends itself to an uncivilized version of nature, and in deference to his readers' sensibilities attempted to palliate the ugliness, for instance, denying in the *Origin* ([1859] 1964) that animals live in fear. Whether nature is cruel or humane, such language is not literal, but an application of an affective perspective.

Contemporary scientific writers do not shy away from introducing figuration to persuade their readers. The earth, L. Thomas writes, is, in a way, a living cell. The evolution of species, according to S. J. Gould, rolls forward fitfully more as a polyhedron might roll than as a smooth ball would. Genes are called "selfish" by sociobiologists. In these cases, the writers intend to lay a sort of metaphorical cover over their descriptions, telling us how to make sense of them. Such figures, the presumption is, could be cashed in for something more precise and controlled, something literal. A gene is not literally selfish, perhaps, but selfishness stands as a kind of folksy shorthand for the fact that some genetic material that seems not to serve the host organism nevertheless gets passed on to the next generation. This is how I read the stance of R. Dawkins's *The Selfish Gene* (1976). Once introduced, however, the rhetoric takes on a life of its own.

As the notion of the "selfish gene" gains currency, it starts to look like a literal claim—as though genes had the personality attribute of selfishness. If one follows Burke (1969), this is not surprising, as one would expect to find the properties of "agents" at work in any account of nature. If it is not Mother Nature or Darwin's selective breeder, then it is perhaps the organism plotting its survival, or even the gene itself, acting with purpose and foresight, executing strategies, recognizing its kindred genes in others and responding in solidarity.

The more one probes, the more difficult it seems to draw any distinct line between metaphorical and literal uses of language. We know that many terms originating as clear instances of metaphors in time cease to be recognized as such. The "leg" of a table might have originally been a metaphorical designation, drawing a comparison to the leg of a biological organism, but we no longer think of it in that way. The term has become literal. Yet the associated imagery of a biological body is there, available for rhetorical use, and perhaps in any case its resonances work on us in ways that escape attention. In the obvious cases, however, where everyone would agree that terms are being used metaphorically, we usually allow poetic license so that the metaphor can play heuristically without being held to rigorous referential criteria.

If metaphors require a certain poetic license in discourse, not a relation of referential fixity, one might think that the highly controlled language of science would not find an easy coexistence with metaphorical language. Accordingly, one might think metaphorical language is a less rigorous idiom appropriate to popularization, but not so appropriate to actual science. Arguably, instances such as Darwin's "selective breeder" are used for the benefit of the popular audience who might not easily grasp the technical arguments. In that case, metaphors would be the bridging devices between technical talk and a more widely shared medium. There seems little question that this is an important means of making the connection between the popular and the technical; imagery of the everyday and familiar are used as resources to help the reader map out an unfamiliar territory.

Yet a look at the discourse of biology belies any simple dichotomy between the technical and the popular. Here the figurative and technical forms of language seem to coexist side-by-side within biology. Just as ordinary people use metaphors as bridges between the familiar and the unfamiliar, and to persuade, so do scientists. Pragmatically, metaphors help with the situated management and mismanagement of meaning. In the discourses of science, they serve to undergird both comprehension and communication, but they also produce their share of misunderstanding and miscommunication.

The problem of misunderstanding occurs partly because metaphorical language does not serve only semantically. Is it true to say that the gene is selfish? Is it false? The answer must be ambiguous, and science is uncomfortable with ambiguity. Within science, metaphors pose the problem of lacking explicit criteria for giving them theoretical weight or relevance. Ironically, this is probably the very characteristic that makes them so easily transportable across disciplinary bounds and adaptable to different contexts. If a discourse of tight technical standards and regulated contexts is to enter and influence other discourses, then perhaps it must be appropriated somewhat loosely, even figuratively.

Population ecologists must assume that microscopic genetic processes underlie what they observe, even if their references to genes will be only crude placeholders in their vocabulary. But a further assumption—and this would be crucial to a scientific attitude—is that the loose references could in principle be pinned down in some measurable way, according to established criteria. In other words, the heuristic force of the metaphor would be only one side of its use to be rounded out in principle by some criteria for its appropriate application in different contexts. What can happen, however, is that the force of a metaphor will precede any controlling criteria (see S. Toulmin's [1958] notion of force and criteria as applied to modal terms in argument). The conjunction of force and criteria established in one scientific context can thus be lost in another.

The suggestiveness of an image can make it tempting to use in a variety of contexts, based solely on the cogency and apparent explanatory power of that image; but if science is to be more than just rhetoric, there must also be a form of accountability for images and metaphors that become embedded in theories, specifically an accountability to criteria for determining when and where they should apply. The various branches of the biological sciences, surely, need to have commerce among them, probably much more than now occurs in the highly specialized research establishments. This implies that there must be sufficient common coin for such exchanges, with variable exchange rates.

To talk this way is to invoke the economic metaphor of exchange, which can be useful in exploring the working relationships among the branches of science. This exchange metaphor, however, leaves out the dynamism of rhetoric. One task for a rhetorical perspective is to explain how habits of representation, and the practices with which they are associated, are communicated and made persuasive across disciplinary lines in the absence of a fixed exchange rate. To that extent, for instance, do genetic rhetorics, or "gene talk," elude disciplinary controls as they move into different theoretical environments (Howe and Lyne 1992)? How do changes of context subtly alter the import of these rhetorics? Answers to questions such as these cannot follow from simplistic attempts to discover and root out figurative language in science. The challenge is to better understand how metaphors implicate a broader economy of thought and imagery, not constrained by disciplinary bounds, and how using them may therefore entail trade-offs between insight and control. If differences exist of both syntax and semantics in different regions of discourse, as I assume, it does not necessarily follow that people are stymied by incommensurability. Rather, they may figure out rhetorics, good or bad, that construct a continuity of meaning as they move in and out of these discursive regions.

Scientific Inquiry Moves Forward by Adjusting in Different Regions of Discourse to Demands of Its Various Audiences

It is commonplace to make a rough distinction between science and its popularization. For many purposes, that distinction makes imminently good sense. The problem is that once one begins to inspect the vast collection of practices called scientific, one sees that the distinction hides some rather crucial assumptions about what and where science is. Principally, it suggests that the kind of adjustment made when one goes to an audience only occurs after science is essentially done. The product is crafted, then it is sold. The knowledge is secured, then it is made appealing. Discovery on one side, communication on the other; knowledge here, rhetoric there. No doubt this model works to describe many processes that we value, such as the transmission of scientific theories through textbooks, or the explanation of complex findings to a general public. But many other science-rhetoric relationships matter as well.

The need to be persuasive does not arise just as one goes before a general public. Persuasion is practiced on others after one has a desired conclusion in mind. But to think of persuasion in the conduct of scientific investigation, one must think of persuasion as the dialectical opposite of doubt in the investigator. The shortfall between suspecting something and knowing it provides much of the energy needed for scientific inquiry. The irritation of doubt nettles and goads one toward the fixing of belief, as Peirce calls it. The overcoming of doubt is not, *pace* Descartes, certainty. Rather it is becoming persuaded, reaching adherence in the face of something less than perfect certitude. Having the means to reproduce that adherence in others is the difference between having a hunch and having the proof. The methods and procedures used to secure conclusions can be viewed as the machinery of persuasion specific to the scientific practice in question.

The practice of science also occurs over networks of people who share highly refined methods of communication. But when theories in one region of scientific practice catch the attention of those working in other regions, who may not share the same body of specialized knowledge, then we have something at least analogous to popularization. That is, the arguments are received within a somewhat different idiom from the one in which they originate. Perhaps an infinitely shaded continuum of sameness-difference describes how the assumption of those taking up scientific arguments, information, or models differ from those of the originators of knowledge. And there might be a variety of ways of breaking up that continuum. Fuller (1988), for instance, distinguishes the originators of scientific knowledge from those who apply it, and again from those who benefit from it. Another way is to think more in terms of disciplinary practices and assumptions and the ways that knowledge claims cut across them. Thus, one can distinguish between those audiences circumscribed by a set of disciplinary assumptions and those who, although scientists, work within a different disciplinary matrix.

This would lead us to mark the difference between intradisciplinary and interdisciplinary scientific discourse. Sociobiologists, for instance, often receive and incorporate knowledge from different disciplinary specialties and thus help mold an interdisciplinary audience. When developments in the sciences go to general circulation newspapers and magazines they reach a nondisciplinary audience. This audience's ability to apply routine knowledge of statistics, or of other tools widely used in the sciences, make it more susceptible to broad imagistic or argumentative strategies. And yet, again, this dynamic of popularization is more a difference in degree than in kind from what occurs whenever knowledge claims travel to different audiences. Given a high degree of specialization, a field can comprise many different audiences.

To the rhetorician, the differences among audiences are as instructive as their commonalities—one should lose sight of neither. Having chased down differences among audiences and discourse practices, one also needs to think about how they relate to one another—what sorts of unities there are among the audiences for scientific discourse. If scientific understanding does move toward broader orbits of acceptance, as successful science tends to, then it has to bring broader audiences under its sway. We might well conclude that a notion such as that of the "gene" might in one sense mean something different to different audiences: different to the laity than to scientists, and still different to scientists working in different regions of discourse. These differences may arise in part from varying degrees of specificity and theoretical embeddedness. They might also arise from inadequate information and misunderstanding—there is, after all, no reason to relinquish the category of error in order to acknowledge different ways of understanding.

Science Enters Public Attention Through the Rhetorical Performance of Testifying Experts

If one thinks of science as a project authorized by the broader community because it is expected to benefit that community, then one is pressed to think of how a scientific mode of understanding becomes absorbed by more general modes of understanding. The question that Peirce pressed when entertaining the notion of a Speculative Rhetoric that would embrace the sciences was a broader, community-oriented question. How do scientific theories and practices become embedded in the thoughts and habits of a scientific community? How are they communicated and appreciated? These are not merely sociological issues, if one takes Peirce seriously, but part of the broader purview of scientific rationality. That larger project of communication and, indeed, persuasion, is as much an interest of "the scientific intelligence" as is the laboratory experiment. Scientific rationality requires syntactic rigors, semantic ingenuity and accuracy, and rhetorical efficacy. Otherwise it dies on the vine, never yielding that full inferential metaboly that a pragmatic theory seeks. Being persuaded, so often the benchmark of rhetorical success, is the very quest of the curious but skeptical mind, that is, of the scientific intelligence. Being *falsely persuaded* is its bane, the very danger Plato feared.

And what are the criteria for distinguishing and isolating false persuasion? The answer, all attempts to the contrary, is that no general answer is possible. That which "justifies" persuasion is as complex and nuanced as the content and circumstances of any given domain of investigation. And it is even more so, because the ways and means of persuasion are not just the domain-specific ones. Different audiences are persuaded by different evidence and argument. This is not to embrace a wide open populism that would make the opinions of the average person the equal of the expert. For those questions that are domain specific, the person best positioned to be persuaded is the person who knows the most about the given subject—knows the false leads that have tripped up others, as well as the fuller story behind any snippet that the novice might pick up. So, up to a point, what persuades those best in a position to know is what we should value, even if it is necessary continually to debate the question of who, in any given matter, is in the best position to know.

Aristotle argued that every rhetorical argument may be judged in respect to *ethos* and *pathos*, in addition to *logos*. That is, the credibility of the speaker, as a social construction, weighs heavily in our evaluation of the argument, as does the ability to connect to the affective dimensions of the situation for the audience. In a well-gauged rhetorical performance, these complement the reasoning being advanced, making it compelling. This gives the nonscientific audience grounds for receiving critically or approvingly the claims of experts. One must respect the suspicions that sometimes arise when expert testimony is introduced to sway public opinion. When experts speak in a public forum—the courts, before congress, on television—the very fact that they are in such a forum usually indicates that the matters they are addressing have more than a technical dimension.

General audiences accustomed to seeing experts testifying on opposite sides of a question have learned to be wary. They know that few questions that come before them, where their own assent is being called for, are likely to be purely technical questions (McGee and Lyne 1987). Expertise is often in the service of money, or some other external motivation. The expert must therefore be viewed with a critical eye. In fact, it can be argued, a part of the necessary preparation for receiving expertise is to apply a touch of rhetorical criticism, "Why is this person in this situation trying to get me to believe just this?" Knowledge is disseminated by testimony, and comes to reside in the larger community primarily by that means. Such testimony needs to be viewed and evaluated not just as an epistemic but also as a rhetorical performance.

A scientific rhetoric, like any other, must be persuasive to succeed. The burden inherent in making a scientific theory persuasive is that it has to be more than appealing on its own terms, but regarded as consistent with other things known. The audiences of scientists and nonscientists who receive scientific discourse are, from a rhetorical perspective, users of meaning, who place models and theories into various relationships, using various strategies. Moreover, they are motivated by various external factors, and within the framework of contested intercontextual understandings. Meaning holds together in the way it comes to bear on argument and persuasion.

Meaning Develops and Extends Pragmatically

The philosophy of science, by and large, has not been especially successful in educating scientists and nonscientists to the role and nature of language in scientific investigation. Theories based more on idealizations than on historical realities have been at the center of the philosophy of science since long before the logical positivists appeared on the scene. The great debate about the discontinuity of paradigms, while it has led to an expanded interest in models and metaphors, as well as to debates about the possibility and commensurability of reference, has been conducted with relative insensitivity to the powers and mechanisms of language.

The notion of the "gene" is fertile for considering how the term was meant in different contexts. In his history of the scientific notion of the gene, E. A. Carlson (1966) wrote the following:

The gene has been considered to be an undefined unit, a unit character, a unit factor, a factor, an abstract point of a recombination map, a three-dimensional segment of an anaphase chromosome, a linear segment of an interphase chromosome, a sac of genomeres, a series of linear sub-gene", a spherical unit defined by target theory, a dynamic functional quantity of one specific unit, a pseudoallele, a specific chromosome segment subject to position effect, a rearrangement within a continuous chromosome molecule, a cistron within which fine structure can be demonstrated, and a linear segment of nucleic acid specifying a structural of regulatory product. (P. 259)

Given this diversity in how the term "gene" was meant, it poses a continuity problem in two respects: the continuity of meaning necessary to communication, and the continuity of reference necessary to an accumulation of knowledge over time. How, one might ask, has a notion whose definition has been specified in so many different ways carried a sufficiently stable meaning to make scientific communication within, across, and beyond the disciplines possible?

The problem goes beyond one of meaning shifts. The fact is that the classical concept of the gene, as Kitcher (1985) notes, was not originally well defined; but Kitcher finds this no problem for the practicing scientist in a newly developing field: "This and numerous other instances—'atom,' 'electron,' species,' 'energy'—testify to the fact that new sciences need not undertake the difficult task of providing neat definitions by focusing on central cases, and later investigators can try to resolve the worries about how exactly the usage of the term is to be extended" (p. 343). Rigorous and exact definitions come relatively late in the development of a science. This "pragmatic approach" to scientific theorizing, Kitcher argues, is to work from strong examples toward theoretical and linguistic clarity. But what, in the meantime, of communication? And what, precisely, about the strong cases is preserved from the state in which definitional clarity has not yet been achieved? The pragmatic way in which these things actually unfold, if Kitcher is right, would seem to raise fundamental questions about the role and nature of language in scientific investigation. And it would make the possibility of scientific communication, and hence meaningful argument, especially problematic. For in the absence of clear and fixed meanings, how well can language communicate? How are we to know that a term means the same thing whenever it is introduced? How can such imprecision fail to undercut scientific measurements and the coordination of activities at different locations?

I believe the two issues, the accumulation of knowledge and the possibility of communication, are intimately connected, because each presupposes continuity in what might broadly be called meaning, and more specifically in the case of science, the possibility of shared reference. What sort of continuity in meaning and reference obtains across changing scientific theories is very much up for grabs among philosophers of science. The third continuity that is implicated by the other two is the continuity between the discourses of science and those of society.

The difference between theories stressing continuity and discontinuity in scientific conceptual change has pivoted on certain received theories of language, which are based largely on idealizations (Burian 1985). To simplify, arguments for discontinuous and incommensurable theoretical developments have been influenced by theories of reference in the tradition of Frege. The reference for any term, on this view, is whatever fits the description the speaker is willing to give to back up the particular usage, that is, the "sense" of the term. In other words, the sense of a term is presumed fixed.

Arguments for continuity, on the other hand, have been based on what is known as a causal theory of reference, based on work by Donnellan, Kripke, Putnam, and others. This theory, the philosophers' leading alternative to Fregean theory, holds that the import of terms traces back to original uses. But the apparent historical sensitivity of this view is only superficially so. Putnam (1975) and Donnellan (1974) advanced versions of a "baptism" theory, which holds that some original link between a term and its reference is forged, and that this link is transmitted over time. A concern with historical change and transformation, the actual dynamics of language, are not a part of the theory. This theory privileges origins, and subsequent relations to origins, in its theory of meaning, a rhetoric of what Burke (1969) would call substantiation by origins, rather than, say, familial relations, or any of Burke's other means of "substantiation."

Each theory is in its own way "closed": In the first case, "sense" is taken as a quasi-platonic boundary on reference, set by either a paradigm or something more transcendent; in the second, there is a causal chain leading from the original designated meaning to present usage, with continuity of reference guaranteed. "Closure of reference," as Burian observes, makes conceptual change into an all-or-nothing matter. Both general theories try to make reference connect in some reliable way to realities past and present. The way that the language theory is connected with the problem of conceptual change in science is much influenced by an obsession with vindicating realism. If knowledge about real things is to be progressively obtained, the reasoning goes, then the flux of language must somehow be arrested. The above approaches represent two strategies of arresting the flux, the first by trying to preserve the entities referred to by earlier scientists, the other by trying to preserve the evidences that made them speak as they did (Fuller 1988, 66). Each seems to assume that that language works in only one way to secure reference in science, and furthermore, that scientific meaning must be fully determined by some combination of sense and reference.

Burian (1985) and Kitcher (1978) have invoked the notion of "referential capacity," a notion of openendedness in the ability of terms of refer over time, as a basis for reexamining the prevalent language theories. I find this notion useful, compatible with the way rhetoricians tend to see language. Kitcher argues persuasively that philosophy of science should abandon its traditional assumption that one can recon-

The indexical sign identifies something, not by resembling or characterizing it, but by some real, nonlinguistic connection to what is signified. Thus the pointed finger singles out something only if that something is in the right location. One can index something but misidentify it. Scientists can successfully make reference to a gene on the basis of an effect wrongly attributed to that gene. Thus a gene associated with alcoholism might be singled out for purposes of communication, even if it turns out not to be causally connected with alcoholism. This is essentially the insight behind the causal theory of reference, which seeks to reconstruct what someone had in mind in identifying certain signifiers, but to postpone judgement about whether things so indexed are identified correctly from a scientific point of view. The indexical component of meaning depends on something being identified, singled out, located, connected to, outside of the talk itself, so of course it has a special pertinence to the goals of science. And it can resist theoretical fixity, or even disrupt symbolization. This is possible only because users of theory are not fully ensnared in the theoretical net.

The third term in the typology, the iconic, may account for much of what is not adequately accounted for by semantic theories because it is determined pragmatically. An iconic sign carries a significance because the physical character of the sign itself carries information. A map is a good example. Usually the iconic sign is said to "resemble" that which it represents. At the most general level, it would be more accurate to say that the iconic sign characterizes something within the frame of some use. Imagery, graphic depictions, imitations of sounds, and many gestures function iconically. One would not have to know the language to read them. International traffic markers are iconic, as are pictures for preliterate children who do not know the language.

Interestingly, the iconic aspect of signification is capable of being highly indeterminate in its semantic content while pragmatically precise. That is, a picture is not only worth a thousand words—it may yield a thousand different words from each person asked to translate it into verbally determinate meaning. Its semantic imprecision is why it is often overlooked in talk about science. Yet insofar as meaning is carried iconically, it has the capacity to yield, not only new insight, but actually new information to its users. What is learned from it, even in the absence of further information, can be more than was encoded. Thus, given a few shapes and angles the geometer is set for business discovering relationships among these imaged entities. One can adduce from a map that Pittsburgh is closer to Princeton than to San Diego although no one had to mark this specific relationship on the map.

What can be said about the different historical uses of the word "gene" is not just that they index one or many different things, which they may, or that they provide sufficiently precise linguistic conventions to support theory, but that there is a vague set of imagery, social in origins, that hovers over these uses and provides a heuristic for thinking about them and for thinking about them as part of the same thing. The vague imagery is by nature unbounded and free to cross all disciplinary boundaries. It permits social scientists to believe that they somehow talk about "the same thing" as geneticists, and for the various branches of biology who use genetic notions to believe that in some sense they are on to the same thing (Howe and Lyne 1992).

In summary, the iconic characterizes, the indexical identifies (by localizing rather than characterizing), and the symbolic correlates within a pragmatic flow. Communication depends on the triangulation among the three aspects, even when (perhaps especially when) there is a deficiency in the other two. But the process is never under the full control of theory because the *users* are not. I can see what the social scientists are talking about when they refer to "traits," even if I do not believe their characterization (icons) or their theories (symbols). Some uses of biological terms in social contexts, even by scientists, rely on iconic rather than either symbolic or indexical signifiers. In other words, one can "envision" genes, natural selection fitness, and so on, without even adducing systematic evidences of them (ibid.).

Rhetoric Operates within the Context of Practices, Not a Closed Universe of Words

We know that the practice of science does not occur within a vacuum. Discourse is embedded in practices, and vice versa. Practicing scientists gets a sense of what is important and worth pursuing from broader social values and, conversely, we expect scientists, whether supported by public or private funding, to bring something of value back to the broader society. This mutual implication of values alone would prevent scientific rationality from being a closed system. The picture that I would want to paint of science is one of a cognitive continuum, no less than one of a practical, rhetorical, aesthetic, or ethical continuum—in which knowledge and techniques learned in general life transfer to the more specific contexts of science. Puzzle-solving and pattern recognition skills learned from amateur sleuthing, playing computer games, and figuring out what caused one's favorite house plant to die are patterns that recur in scientific investigation. And, I suggest, tropes and figures applicable in broader life seem just as obviously—once we take off the doctrinal blinders—to operate within science as well. Obviously, scientists attempt to discipline these more general habits, as well as to eliminate the bad ones. That discipline, thankfully, is imperfect.

The reasoning about how genes operate did not originate full-blown when we learned the mechanics of the double helix. Rather, existing patterns moved in around the new knowledge to give it a context and efficacy in evolutionary explanation. Terms deeply embedded both in science and in culture—terms such as selection, fitness, and adaptation—continued to carry the weight of accumulated meaning and association. No technical specification cleanses these terms of those associations.

Nor do practices reduce to theories. Theoretically, each level of biological science should have its place in a hypothetical grand reduction, in which every level of explanation is consistent with every other. In practice, such a consistency is not required. Molecular geneticists can and sometimes do doubt the prevailing theories of evolution, for instance. The undigested parts of the collective scientific enterprise sustain their own activity, consistent or not with the next level up or down the chain of reduction. Practitioners who know their jobs well may know nothing about what other practitioners do, still less about what others see them as doing. Viewed in terms of meanings, this suggests a great deal of nontransparency. It also suggests a great deal of de facto inconsistency. To understand science as discourse requires some sense of the macro- and microlevels of discursive practice in which any given discourse is embedded, because these levels interanimate one another.

To speak of levels of discourse suggests hierarchy, and especially in the case of science, the promise of reduction. To speak of regions of discourse, on the other hand, is to suggest areas of application that may be remote, conceptually or practically, from one another. The highly specialized nature of organized science produces regions of discourse that can be isolated from one another even if carried out in close geographical proximity. One writes to fellow specialists who read the same journals, not necessarily for the colleagues down the hall. Practices may isolate as well as consolidate.

Rhetorics Take on a Persuasive and Simplifying Force of Their Own

Rhetoricians who make claims about scientific discourse being rhetorical are apt to provoke a fight from those who believe this means some sort of slander on science. Rhetoricians, of course, usually mean it as a compliment. Being rhetorical does not have to come at the expense of being other things. Indeed, it can be bad for rhetoric as well as for science if scientific rhetoric grows into something independent and isolable from scientific practices. Howe and I (1992) have attempted to show how a simplified and streamlined rhetoric can create the illusions of a synthesis from among several distinct traditions in genetic research. The rhetoric floats, so to speak, above the practices and thus drifts somewhat too freely into new territories.

When biometric geneticists apply finely honed equations to specify the meaning of fitness, their arguments cross into other domains where those equations are all but unknown. The terminological configuration of selection-adaptation-fitness traits gallops across a range of different disciplinary and nondisciplinary contexts, carrying what can meaningfully be called rhetorical force, independent of disciplinary controls. If one looks at the way the icon of "fitness" is used across the discourses of genetics, evolution theory, and sociobiology, for instance, it appears that the suggestive power of the terminology is part of what connects the different levels and regions of discourse.

The notion of natural "selection" carries residues of the everyday usage, that is, connotations of choice in selecting, which is closely associated with purposive action. Darwin embellished these connotations with overtones of an all-wise Nature, culling as necessary for the greater good. Social Darwinists elided the job of nature and the job of social institutions in selecting the fittest. Now sociobiologists have the genes colluding with cultural practices to select the most promising behaviors—promising, that is, from the viewpoint of the genes, who struggle to insure their own survival. This displays a pattern (symbol, in the sense of being conventionalized) invented in the nineteenth century, putatively indexed to recent knowledge about the genetic stuff of life. The implication is that we have, in the science of genetics, a new basis for doing Darwin right, so to speak. The primary mechanism linking sociobiology to actual genetic knowledge appears to be analogy, but there is really no disciplinary warrant or evidentiary link to the sciences of genetics. There are many rhetorical links.

Applying reasoning patterns developed in one area to other areas is common, and it is one of the ways that understanding grows. E. O. Wilson's *On Human Nature* (1978), which has had a pervasive impact in the universities—in places such as philosophy and sociology departments—is a bold set of speculations about genetic origins of behavior which seems to rely very little on what is known of genetics and a great deal on analogies (Lyne and Howe 1990). In his and other speculative work coming from sociobiology, there is a use of what can be called black box genetics, which posits genes for "traits" such as aggressiveness and altruism. If most sociobiologists prefer to steer clear of the "trait" of criminality, it is more likely a reflection of their political sensibilities than of their explanatory apparatus. (Predictably, others are increasingly talking about supposed genetic bases of criminality.) Because it is not clear what sort of disciplinary constraints may obtain when sociobiology forages for support among the various fields of population genetics, ethology, anthropology, economics, and psychology, among others, it can become a simplifying style of explanation for anyone inclined to use it.

Part of the success of sociobiology derives from its willingness to use a rhetoric that suggests genetic determinism, while making no direct claims to that effect. References such as "our hidden masters, the genes" (Wilson 1978, 4) do the work of isolating the rhetoric from any robust involvement with scientific inquiry while exercising a persuasive force on readers. The genetic *perspective* is thereby carried beyond what the sciences of genetics have provided. Genes as heuristic becomes genes as warrant. A biological perspective becomes a bio-logical claim. Obviously, injecting some reminders into popular discourse that we have a biological nature is not in itself something to be faulted. To argue that cultural beings are guided by their genes in ways that are homologous to ant societies, however, sacrifices balance and proportion to emphasis.

Those of us who wish to celebrate rhetoric need also to assemble reminders of rhetoric gone bad. Rather than dismiss the rhetorical forces that combine to make a scientific argument persuasive, one needs to take them seriously and engage them in critique. This would entail an examination of rhetoric figures to see not just if they work, but from where they may draw their rhetorical efficacy. This would be part of a general critical consideration of language and imagery in scientific discourse.

NOTE

Portions of this essay were presented at the 1989 Summer Conference on Argumentation at Alta, Utah.

REFERENCES

Apel, K. O. 1981. *Charles S. Peirce: From Pragmatism to Pragmaticism.* Translated by J. M. Krois. Amherst, Mass.: University of Massachusetts Press.
Burian, R. 1985. "On Conceptual Change in Biology: The Case of the Gene." In D. Depew and B. Weber, eds., *Evolution at a Crossroads.* Cambridge, Mass.: MIT Press, pp. 21–42.
Burke, K. 1969. *A Grammar of Motives.* Berkeley and Los Angeles: University of California Press.
Carlson, E. 1966. *The Gene: A Critical History.* Philadelphia: Saunders.
Darwin, C. [1859] 1964. *On the Origin of Species: A Facsimile of the First Edition.* Cambridge, Mass.: Harvard University Press.
Dawkins, R. 1976. *The Selfish Gene.* Oxford: Oxford University Press.
Donnellan, K. 1974. "Speaking of Nothing." *Philosophical Review 83:* 3–31.
Eco, U. 1976. *A Theory of Semiotics.* Bloomington: Indiana University Press.
Fuller, S. 1988. *Social Epistemology.* Bloomington: Indiana University Press.
Howe, H., and J. Lyne. 1992. "Gene Talk in Sociobiology." *Social Epistemology* 6: 109–63.
Joravsky, D. 1970. *The Lysenko Affair.* Cambridge, Mass.: Harvard University Press.
Kitcher, P. 1978. "Theories, Theorists and Theoretical Change." *The Philosophical Review 87:* 519–47.
———. 1985. *Vaulting Ambition: Sociobiology and the Quest for Human Nature.* Cambridge, Mass.: MIT Press.
Lyne, J. 1980. "Rhetoric and Semiotics in C. S. Peirce." *Quarterly Journal of Speech 66:* 155–68.
Lyne, J., and H. Howe. 1990. "The Rhetoric of Expertise: E. O. Wilson and Sociobiology." *Quarterly Journal of Speech 76:* 134–51.
McGee, M., and J. Lyne. 1987. "What are Nice Folks Like You Doing in a Place Like This? Some Entailments of Treating Epistemic Claims Rhetorically." In J. S. Nelson, A. Megill, and D. N. McCloskey, eds., *Rhetoric of the Human Sciences.* Madison: University of Wisconsin Press.

Peirce, C. S. 1931–1958. *The Collected Papers of Charles Sanders Peirce*. Vols. 1–6 edited by C. Hartshorne and P. Weiss. Vols. 7–8 edited by A. W. Burks. Cambridge, Mass.: Harvard University Press.

Putnam, H. 1975. *Philosophical Papers*. Vol. 2, *Mind, Language, and Reality*. Cambridge, England: Cambridge University Press.

Shapiro, M. 1983. *The Sense of Grammar: Language as Semeiotic*. Bloomington: Indiana University Press.

Suppe, F. 1989. *The Semantic Conception of Theories and Scientific Realism*. Urbana: University of Illinois Press.

Thompson, P. 1989. *The Structure of Biological Theories*. Albany: SUNY Press.

Toulmin, S. 1958. *The Uses of Argument*. Cambridge, England: Cambridge University Press.

Wilson, E. O. 1978. *On Human Nature*. Cambridge, Mass.: Harvard University Press.

Comment: Metaphors, Rhetoric, and Science

Michael Bradie

Department of Philosophy, Bowling Green State University

I am sympathetic to the general drift of J. Lyne's essay. In particular, I agree with Lyne on the importance of metaphors for science and the need for rhetorical analyses of science although I have a slightly different story about the relationship between metaphors, rhetoric, and science. I will begin with a sketch of the role of metaphors and rhetoric in science, then I will challenge some of the specifics of Lyne's analysis. Lyne suggests that a rhetorical approach to science is superior to a syntactic or semantic approach and, though in some respects that is undoubtedly true, I suggest in others it is not. I take exception to his analysis of the sociobiology case and will present an alternative and fairer assessment of the situation. Finally, I question some of the details and promise of Lyne's approach to how a rhetorical analysis of science should proceed.

The Metaphorical Structure of Science

The role of metaphors, models, and analogies in science is complex. For much of this century these tools were relegated to heuristic roles only. Since the decline of positivism in the 1960s, new studies suggest a more pervasive and central role for such tools in all aspects of scientific activity. This is not the place for a detailed analysis. I only sketch the rhetorical dimensions of the role of models and metaphors in science. First, let me distinguish between rhetoric in the wide and narrow senses. In the wide sense, rhetoric is concerned with all linguistic devices used to communicate and persuade. Metaphors (and I will not distinguish between metaphors, analogies, and models here although much has and can be said about their similarities and differences) are one trope among many. In this sense, all metaphorical applications are rhetorical. However, I think it is important to distinguish, under this wide umbrella, between rhetorical (in a narrow sense) uses of metaphor in science and heuristic and cognitive uses. In the narrow sense, rhetorical uses of metaphor in science include the communication of scientific knowledge both technical and popular and the use of scientific models to persuade and mold opinions in the sociopolitical domain. The heuristic use of metaphor includes the use of models and analogies of various sorts to suggest lines of theoretical research or the shape of new theories based on old familiar notions. Finally, the cognitive use of metaphor includes the explanation of empirical phenomena through the identification of empirical systems with theoretical models (see Bradie 1984, 1987 for further details). Sorting out the implications of this interconnected set of uses is a complex matter. The endorsement of the centrality of metaphors in science should not be construed as an endorsement of the centrality of merely persuasive devices in science.

The Relation of Rhetorical to Syntactic and Semantic Analyses of Science

Lyne suggests that a rhetorical analysis of science is to be preferred to the classical positivistic analyses and to the more recent analyses of the semantic approach. While I agree that these approaches are incomplete, I see rhetorical analyses as complementing rather than replacing them.

We may consider a phenomenon to be analyzed which I will call scientific practice. This is to include the usual activities of observing, setting up experiments, inventing and perfecting devices, testing, predicting, explaining, describing, communicating, teaching, lobbying for funds, and the rest. Science, so circumscribed, is a complex phenomenon amenable to any number of analytic approaches, for example, psychological, sociological, historical, and philosophical. Like any complex phenomenon, we approach it by producing a model which highlights the features we may be interested in and attempt to understand the phenomenon through reconstructing it in terms of the model. Borrowing a phrase from the positivists and expanding its range of application, we may call these models "rational reconstructions."

Although recent reassessments of the positivists should caution us against glibly assuming a monolithic point of view among them, at least those whose ideas are embodied in what came to be known as "the Received View" approached the analysis of science from a singular and somewhat narrow perspective. They focused their attention on the logic of the relationship between theory and evidence and the question of the validation of scientific claims. In pursuit of this end, they lobbied for a particular method of reconstructing scientific theories which construed them to be axiomatic systems à la Euclidean geometry. Without detailing the rise and fall of this approach, by the 1960s, when the program was beginning to lose its dominant position among philosophers of science, two points had emerged. First, the program did not work. The axiomatization along strict lines was more difficult than the positivists had assumed. No "living" science could be properly axiomatized and even if it could there did not seem to be any advantage to so doing. Second, the program was incomplete. Many of the complex phenomena included in scientific practice were not dealt with by the positivist approach. Critics complained that scientists rarely, if ever, behaved as their positivist reconstructions did; but the official response was that the positivist model was not intended to reconstruct the entire practice of scientists but only the logic of a certain body of scientific claims.

In the 1960s, new approaches, which focused on aspects of scientific practice ignored by the positivist approach, began to gain ascendancy. A leading figure in this realignment was T. Kuhn (1970). His work created many waves and ripples, one of which contributed to the development of the other standard approach Lyne rejected: the semantic approach. This is a complicated story which involves contributions by Sneed (1971), Stegmuller (1976, 1979), Suppe (1977), van Fraassen (1970), and others. Several versions of the approach were first applied to theories in physics and then, more recently, to biology, by advocates such as Lloyd (1988) and Thompson (1989). The details and the differences need not concern us here. The important point is that the semantic approach is fundamentally designed as a substitute for the failed axiomatic approach of the Received View. As such, its primary focus is the same as that of the positivists. Scientific theories are construed as collections of models with applications, and explanations involve construing empirical phenomena in the light of these applications and models. In contrast to the Received View, the precise syntactic details of the relationship between theory, descriptions of phenomena, and evidence are unimportant. The jury is still out with respect to this approach, but, again, two points are clear. First, the program is largely untested. It is unclear whether, in the long run, it will prove to be substantially easier to construe theories as collections of models than it was to construe them as axiomatic systems. Second, the approach is incomplete. Again, like the positivistic analyses, the semantic analyses do not pretend to be a complete reconstruction of scientific practice.

Lyne's complaints about these classical approaches seem to run these two distinct points together. He argues that both reconstructive approaches ignore, among other things, the dynamics of (1) interscientific communication between differing professional groups (e.g., within genetics), and (2) interpractice communication between scientific practice as knowledge and understanding producer and other social practices whose focus is on action rather than understanding per se. He contends, in effect, that both classical approaches fail to appreciate the extent to which the medium (language and the use of metaphors and analogies) modulates the message (knowledge content). Undoubtedly, a role for rhetoric exists here. It fills a gap in our understanding of science left by both the positivistic and semantic approaches. This, however, is not a reflection on the wrongness of the classical approaches but merely of their (self-admitted) incompleteness. I am more sympathetic to viewing rhetorical analyses of the sort proposed by Lyne as complementing the efforts of the positivists or semantic theorists than I am to seeing them as substitutes for them. Whether the classical analyses are correct or useful for the restricted domain they intended to explicate is a separate issue.

Human Sociobiology: Sense or Nonsense

As an example of the confusions that can reign if the rhetorical dimension of science is ignored, Lyne offers the case of human sociobiology. He sees this "discipline" as riddled with half-truths, unsupported claims, and driven by political agendas. He is not alone in this assessment. According to Lyne, sociobiology has "no disciplinary warrant or evidentiary link to the science of genetics." He sees sociobiology as using genes and genetics as a heuristic model for human behavior and then illegitimately claiming genetics as a warrant for sociobiological claims. Obviously, merely invoking some model as a heuristic tool is not in itself a warrant for the validity of the model. All such claims of warrant of course need to be checked. But, I think it is wrong to suggest that sociobiological analyses of human behavior are wrong in principle.

E. O. Wilson (1975) claimed that the extension of evolutionary biology to an analysis of human behavior was the natural completion of the modern synthesis. He has been much criticized for this and no doubt much that has been produced under the banner of human sociobiology was half-baked and ill-supported. But Wilson's basic insight is sound. Human beings are animals who share an evolutionary heritage with other evolved living creatures. Social behaviors are phenotypic expressions of an inherited genetic endowment. There is strong evidence that the social behaviors of nonhuman animals can be profitably analyzed from the standpoint of evolutionary biology. There is no reason in principle why human social behavior should be an exception. Of course, the problems of formulating, testing, and evaluating claims about the evolutionary or biological basis of human behaviors are formidable. But these are, I think, mainly technical difficulties. One set of problems arises from difficulties at the psychological and sociological level in identifying proper "traits" for analysis. Another set arises from difficulties in separating narrow biological factors (i.e., genetic factors) from cultural and environmental factors. These difficulties are not so great for the sociobiological study of nonhuman animals because experimental protocols can be established to control for these diverse influences. However, even if we knew how to factor human behaviors at the psychosocial level, ethical constraints make experimentation with human subjects difficult if not impossible.

Given the potential for misuse of claims about the biological bases of human behaviors, some of the early critics suggested that sociobiological research into human behavior should be abandoned and that resources not be directed toward these efforts. These are legitimate concerns and the proper distribution of research funds and effort is surely a sociopolitical question with strong rhetorical overtones. Still we must distinguish between the potential validity of a line of research from questions about the difficulties of pursuing it or the wisdom of funding or supporting it. Rhetorical analyses of science which blur these differences do not serve us well.

Peirce's Triadic Analysis

As a potential substitute for syntactic and semantic analyses of science, Lyne proposes to utilize a rhetorical analysis based on Peirce's triadic analysis of terms as icon, symbol, and index. This analysis is supposed to highlight the fact that theoretical concepts, such as "gene," are referentially ambiguous and equivocal. Lyne sees the referential ambiguity and equivocal nature of such terms as central to making credible the illegitimate extension of genetic models to the explanation of human behavior. I confess I do not get it. Theoretical concepts are often, if not invariably, referentially ambiguous and equivocal as Lyne claims. But in particular cases of allegedly illegitimate deployment, more prosaic accounts of what has gone wrong can be given. For example, sociobiologists often suggest that evolutionary biology can contribute to an analysis of altruism in human beings. Some go so far as to postulate genes for altruism. Appeals are then made to population genetical models of kin selection and the like to show how this works. But no empirical evidence shows the existence of such genes. This does not show, however, that the sociobiological approach, construed as the attempt to isolate and determine the biological or genetical factors contributing to human behavior, is misguided, only that the relationships between the genetic and social level of human behavior are very complicated. Approaching the problem from the social and psychological level, one can analyze human behavior on incest taboos and mating patterns, for example, which do *suggest* the need for a genetical explanation. I do not argue that any such analyses are correct, only that they are not obviously wrong. All of this seems to me to be perfectly straightforward and an everyday occurrence in all sciences where relatively simple models are employed in the attempt to explain complex phenomena. I do not see the need to appeal to Peirce's triadic analysis and, in any case, I am not sure how Lyne sees it as working.

Consider the concept of the gene. It has a symbolic dimension as a designating but arbitrary label. That genes are designated as "genes" is an accident of history. We might just as easily have called them "johns." Second, there is the gene as index. The index of a term covers the collection of rules and criteria by means of which we identify the objects or processes in question and by means of which we use it to refer. Finally, there is gene as icon. The iconic structure of a concept can be identified with the associated models or theories which tell us, in this case, how genes or johns behave. One of the supposed benefits of the analysis is illustrated by the following example. We can use the term "the genetic trait of aggressiveness" to index a behavior and yet be mistaken about whether the trait in question has any significant genetic basis at all. We appear to have imported the concept of "gene" into our discussion of aggressiveness and thereby, perhaps illegitimately, suggested a possible analysis along biological lines. But, even if we are mistaken in our belief about the underlying basis of aggression, we have managed to refer with the term. The net effect is an equivocation. The term "genetic" in "genetic trait of eye color" and in "genetic trait of aggressiveness" does not refer to the same thing.

Let me make two points about this example. First, on the surface, the term "genetic" is not doing the indexing work. That work is being done by the term "aggressiveness." This can be seen by comparing the two terms "the johnetic trait of aggressiveness" and "the genetic trait of perhexnis." In the former case, we have some idea of what this refers to even though no one has any idea what "johns" are or how they behave. In the second case, we know how genes behave but have no clue as to what the term refers. The point is that it is *not* the *theoretical* terms which are doing yeoman service in referring but rather the *non*-theoretical term "aggressiveness." Hence, how a triadic analysis of *theoretical* terms is supposed to be relevant is unclear. On the other hand, my example may be loaded. I introduced a theoretical term, "john," but said nothing about johnetics. If I had, perhaps we would have been led to a conclusion similar to that reached at the end of the previous paragraph, namely, that the term "johnetic" in "the johnetic trait of eye color" and "the johnetic trait of aggressiveness" is equivocal. Then the point seems to be that we should be constantly vigilant against the gratuitous attachment of theoretical terms to referential expressions. Merely because the term appears in such an expression should not mislead us into thinking that a theoretical account of the referent is at hand.

This is a significant and well-taken point, but again I do not see that the triadic analysis is either necessary or superior to other accounts. Similar situations arose in the aftermath of the enormous success of Newtonian mechanics in the eighteenth century. Many saw Newtonian mechanics as the equivalent of a unified field theory, that is, as an explanatory schema that could explain anything and everything. The net effect was that people began to see the "mechanical" aspects of phenomena that no one had suspected to be such at all. Of course, the program was eventually shown to be incomplete and many properties that people thought might be given mechanical explanations turned out to be more complicated. From a "semantic" point of view, the situation is clear. Successful theories generate paradigmatic models which their advocates attempt to extend to cover hitherto unexplained phenomena. Sometimes they succeed. More often they fail. Systems which were claimed to be Newtonian such as light, electricity, magnetism, the atom, the mind, and so on turn out to be impossible to capture by means of Newtonian models. But no complicated Peircean analysis of terms is needed, as far as I can see, to make this point.

Conclusion

Although I am uneasy about and unsympathetic to some of the details of Lyne's proposal, let me end by noting some central areas of agreement. We both stress the importance of models and analogies in all aspects of scientific activity. I am prepared to agree that rhetorical analyses are important tools in the investigation of the communicative and sociopolitical dimensions of science. Claims for support and resources based on designating phenomena as "Newtonian" or "genetic," as the case may be, need to be carefully examined and assessed. To the extent that rhetorical analyses can help us come to reasoned judgements with respect to these matters, they are to be welcomed.

NOTE

This essay was written while I was a visiting fellow at the Center for the Philosophy of Science, University of Pittsburgh. I would like to thank the staff members at the Center for their help.

REFERENCES

Bradie, M. 1984. "The Metaphorical Character of Science." *Philosophia Naturalis* 21: 229–43.

————. 1987. "Metaphors in Science." In *Abstracts of the 8th International Congress of Logic, Methodology and Philosophy of Science, 5:* 000–000.

Kuhn, T. 1970. *The Structure of Scientific Revolutions.* 2nd ed. Chicago: University of Chicago Press.

Lloyd, E. 1988. *The Structure and Confirmation of Evolutionary Theory.* New York: Greenwood Press.

Sneed, J. D. 1971. *The Logical Structure of Mathematical Physics.* Dordrecht: Reidel.

Stegmuller, W. 1976. *The Structure and Dynamics of Theories.* New York: Springer-Verlag.

————. 1979. *The Structuralist View of Theories: A Possible Analogue of the Bourbaki Programme in Physical Science.* New York: Springer-Verlag.

Suppe, F., ed. 1977. *The Structure of Scientific Theories.* Urbana: University of Illinois Press.

Thompson, P. 1989. *The Structure of Biological Theories.* Albany: State University of New York Press.

van Fraassen, B. 1970. "On the Extension of Beth's Semantics to Physical Theories." *Philosophy of Science* 37: 325–39.

Wilson, E. O. 1975. *Sociobiology: The New Synthesis.* Cambridge, Mass.: Belknap Press.

11

Rhetoric, Ideology, and Desire in von Neumann's *Gründlagen*

H. Krips
Department of Communication, University of Pittsburgh

Foucault (1984) argues that the human sciences—medical theories of sexuality, criminology, psychology, and so on—have sustained the modern will to power by turning their knowledges to the ideological and rhetorical tasks of legitimizing contemporary power structures. In a more conservative vein, however, he allows that the hard sciences— biology, physiology, chemistry, and physics—eschew the "stubborn will to nonknowledge" (ibid., 55) which grows out of the will to power. Instead, he says, they "partake of that immense will to knowledge which has sustained the establishment of scientific discourse in the West" (ibid.), and thus are relatively free of the flaws of rhetoric and ideology:

Underlying the difference between the physiology of reproduction and the medical theories of sexuality, we would have to see something other and something more than an uneven scientific development or disparity in the forms of rationality; the one would partake of that immense will to knowledge which has sustained the establishment of scientific discourse in the West, whereas the other would derive from a stubborn will to nonknowledge. (Ibid.)

I argue that such a view is overly optimistic about the resistance hard sciences offer in the face of the contemporary will to power. In particular, I show that a paradigm of hard science, the 1952 English translation of von Neumann's *Mathematische Gründlagen der Quantenmechanik*, repeats contradictory fragments from the ideological field in which it was produced, and thus is ideological in the traditional sense of providing a mirror for independently constituted ideologies.[1] The *Gründlagen* is ideological in Althusser's sense as well, by acting as a site at which subjectivity is created. Theses dual ideological dimensions of the *Gründlagen* also imply its involvement with the rhetorical processes of misrepresentation and the creation of desire.

The rhetorical and ideological dimensions of the *Gründlagen* are most readily detectable at its margins: the translator's preface, footnotes, and the historical introduction. Here the text grapples openly with the rhetorical task of establishing its credibility, a task made more difficult by its insistence upon a degree of mathematical rigor which the audience of physicists considered excessive.

The text's engagement with ideology and rhetoric is not restricted to its margins, however. Even at its center—a plain unvarnished stream of mathematical formulas organized solely by conventions of mathematical proof—the *Gründlagen* is haunted by both rhetoric and ideology. The effaced narratorial voice of the proofs acts as a rhetorical device which spreads authority for the text throughout the scientific community as a whole. The proofs are framed by a narrative of pedagogy in which the narrator sets out to introduce a mathematically rigorous version of quantum theory (QT) to an audience of novices, a narrative convention which plays a key rhetorical role in engaging the text's implied audience. The low visibility of these rhetorical devices may be seen as a further aspect of the text's rhetoric, a way of making it appear more credible by conforming to the antirhetorical canons of scientific writing.

Such rhetorical elements in the *Gründlagen* are not merely a matter of the author's personal style but also a response to forces that arose out of the historically specific institutional context in which the text was originally produced. In what follows I identify those forces and connect them with specific ideological strategies.

My conclusions will leave some readers unimpressed, even disappointed. They will concede the presence of rhetorical and ideological impurities in the narrative conventions which structure and frame the scientific propositions of the *Gründlagen*, but object that this concession does not address the significant question of whether the *Gründlagen*'s ideological and rhetorical impurities are essential to its role as a purveyor of scientific information.

But this objection is not to the point. My deliberations are not intended to bear upon the larger question of whether ideology and rhetoric are essential aspects of the core scientific content of the *Gründlagen*. Instead, I demonstrate the modest claim that, in response to certain forces associated with the institution of science in the 1950s, the scientific heart of the *Gründlagen* was structured in terms of a specific rhetoric and ideology. Broader questions of the nature of the connections between rhetoric, ideology, and science will have to be addressed elsewhere.

The Ideology of Science

The Baconian-Mertonian norms of universalism ("objectivity in assessment procedures" [Merton 1968, 607]), communism ("no scientific private property" [ibid., 610]), disinterestedness (ibid., 612), and organized skepticism (ibid., 614) are not only key elements in science's official discourse about itself, they also influence the practices of scientists. Indeed, insofar as scientific practices involve both a relatively free exchange of information and a certain degree of doxastic autonomy, they provide a lived representation of such norms.

Norms such as these help consolidate science's external power base by portraying the scientist as objective, disinterested and skeptical; and by embodying an ethic which encourages individual scientists to share their symbolic capital with each other, they also contribute to the reproduction of the internal network of scientific exchange. Such effects promote the good of the institution of science at the expense of individual practitioners and the wider public. These benefits for the institution are concealed by the proposition that science works for the good of all rather than to further its own interests, and by a system of honors and awards recognizing individual contributions to science, especially original work. Thus the Baconian-Mertonian norms are ideological in the traditional sense associated with the early Marx (ideology as misrepresentations of an underlying social reality) as well as in Althusser's sense (ideology as a lived representation of imaginary relations).[2]

Eagleton (1982) argues that texts relate to ideology by selectively repeating fragments from the ideological field in which they are produced, a repetition which is not merely a matter of reproduction but rather a reworking that minimizes or at least conceals ideology's contradictory relations with itself. Eagleton reserves the term "ideology of the text" (ibid., 80) for the results of such reworking and concealing.

(Because the contradictory relations between the ideological fragments which fall upon the text are residues of the contradictory relations in the social field that ideology is supposed to conceal or mediate, we can see the task of the ideology of the text as an extension of the general ideological project of concealing or mediating contradictions.) Thus the relation of text to the ideological field mirrors the relation of the ideological to the social: In both cases there is not only a representation but also a reworking or reconstitution that conceals and/or blunts contradictions.

The *Gründlagen* is marked by such "ideology of the text," resulting from its reworking of the contradictions between the official Baconian-Mertonian norms of science and an individualistic ideology that characterizes the broader social milieu in which the *Gründlagen* was produced. The ideological incoherences which survive this reworking become sites at which the text engages readers by and for the production of desire. Thus the *Gründlagen* emerges not only as a vehicle for ideology but also as its source, a crucible for the creation of desire and the imaginary relations which subjects inhabit.

The Narrator

The *Gründlagen's* narration is often couched in first-person plural grammatical form, "We shall omit any discussion of" or "If we can invert R and S then we can invert RS" (1952, 45).[3] In the context of the *Gründlagen* this narratorial "we" represents a genuine plurality, a narratorial consortium, of which the author is only one member. This is achieved by creating a formal distance between narrator and author, "The discussion which is carried out in the following . . . the author owes to conversations with L. Szilard" (ibid., 421, fn. 206) and "A rigorous proof has been given by the author" (p. 143, fn. 91). The narrator is (at least) doubled, to include both a spokesperson for the text as well as the author who is explicitly introduced as a member of a narratorial syndicate. This doubling of the narratorial voice shifts authority from a single source to a plurality. Indeed the text implicates the broader scientific community, or at least a privileged subclass of it, in the process of authorization.

The text also cites other scientific authors—including Dirac, Weyl, Born, Jordan, and Sommerfeld—in an almost personal and invariably collegiate manner, inviting the reader to note the "excellence" or "comprehensiveness" (ibid., vii) of the authors' work, "several excellent treatments of these problems are either in print or in the process of publication" (ibid., vii); a book by Courant and Hilbert is referred to as "exhaustive" (p. 23, fn. 30); and the work of Fredholme (ibid., 23, fn. 30) and Hilbert on integral equations is described as "definitive" (ibid.; Kress and Hodge 1979 offers similar views on the role of the narratorial voice as a carrier of ideology). Such references help create an image of a close-knit community of like-minded scholars, including the narrator, while at the same time legitimating the text by presenting its sources in a good light.

The plural narratorial voice and citations in the *Gründlagen* are not merely rhetorical devices for creating authority. They also present scientists as a community, working together, freely exchanging and using each other's ideas (albeit with suitable acknowledgments) and standing behind the pronouncements of their colleagues. In short, the plural narratorial voice and citations offer a picture of scientific activity as disinterested (impartial) and communal (eschewing private ownership of intellectual property) in conformity with the Baconian-Mertonian normative system.

However, on occasions the flexible "we" narrows its intended reference to the author. For example, in its introduction the text comprehensively condemns not only Dirac but also the bulk of theoretical physicists (ibid., viii, ix). The condemnation begins in a conventionally respectful manner, "Dirac . . . has given a representation of quantum mechanics which is scarcely to be surpassed in brevity and elegance . . ." (ibid., viii) and continues in a suitably modest tone, "It is therefore perhaps fitting to advance a few arguments on behalf of our method, which deviates considerably from that of Dirac. . . ." (ibid.). An overtly critical note then begins to creep into the text, "The method of Dirac . . . (and this is overlooked today . . . because of the clarity and elegance of the theory) in no way satisfies the requirements of mathematical rigour—not even if these are reduced in a natural and proper fashion to the extent common elsewhere in theoretical physics" (ibid., viii–ix).[4] This criticism is then made more specific as Dirac is accused of introducing "'improper' functions with self-contradictory properties. . . . The insertion of such a mathematical 'fiction' is frequently necessary in Dirac's approach" (ibid.; see also pp. 26 fn. 23; 27; 115; 223). The text concedes that "there would be no objection if these concepts . . . were intrinsically necessary for the physical theory" (ibid., ix) but claims that "this is by no means the case" (ibid.). The narrator then presents his own, now clearly meaning the author's own, theory as an instance of the way in which QT can be formulated without such improprieties.

In retrospect, "our" in "our method" is clearly a reference to the author alone, his hitherto "excellent" colleagues having been rudely expelled from their comfortable seats on the editorial board. Thus, in a synecdochical sleight of hand, the narratorial "we" shifts its reference from a whole—an elite within the community of science—to one of its members struggling with a hostile and ignorant audience to gain both credit and credibility for his ideas. By providing a platform from which one scientist personally attacks others in this way, the *Gründlagen* enacts dissent between members of the scientific community, thus displaying, indeed performing, a picture of science as an agonistic field where individuals pursue their own ends in a hostile social environment. This picture of dissent foregrounds the values associated with an ideology of individualism characteristic of the broader society in which science was practiced at the time and place of the book's production—an ideology of open competition, the out-of-hand rejection of competing points of view, ownership of the products of one's own labor with all the personal rights implicit in that concept, and so on.

Translator and Author: A Difficult Relation

The *Gründlagen* includes a translator's preface, "The translated manuscript has been carefully revised by the author so that the ideas expressed in this volume are his rather than those of the translator, and any deviations from the original text are also due to the author" (von Neumann 1952, v). The formal distancing of both author and translator from the narratorial voice suggests an unstable alliance between those who were party to the production of the text. Each, it seems, agrees to coinscribe the narrative on condition that the presence of the other is erased from the narratorial voice.

This same uneasy relation is manifested at the level of the quotation's content. The personal interests of the author dominate the first part of the quotation which affirms that ideas expressed in the text originate exclusively with the author—the ideas expressed in the text are to be "his rather than those of the translator," a theme emphasized by describing variations from the original text as "deviations." By contrast the second part of the quotation indicates that "revisions" have been incorporated during the process of translation; in other words, the author and translator's cooperation resulted in the former's authorizing variations from the original work. And the use of the term "revise" at this point portrays positive relations between scientific colleagues, in particular those between author and translator. It is as if for a brief moment the voice of the cooperative translator breaks through the uneasy narratorial concordance dominated elsewhere by the individualistic author.

In sum, the need for translation to spread the word is implicitly acknowledged, but at the same time dangers of this process are recognized, which require the author to "carefully revise" the translated work and check that no unauthorized "deviations" have been introduced by his careless translator-colleague.

Rival pictures of scientific work are implicit here. On the one hand, the production of scientific text is figured as an active cooperation, while on the other, the author's colleagues and in particular the translator are criticized not only for "deviant" views but also for lack of "care" in reading and writing. This is another variant of the text's central ideological tension: Scientists are members of a cooperative but they are also individuals struggling in a hostile social environment where they have to fight against appropriation of their views. In the next section I show how this tension is repeated within the narrative order of the text. Macherey (1978, 79ff.) calls such a tension that ramifies across and within various levels of the text "a gap."

History and Pedagogy

The text's narrative constructs a narratee who is scientifically knowledgeable both in terms of the advanced literature to which the text confidently refers and those concepts with which familiarity is assumed, such as Lebesgue measure, Cauchy convergence, trigonometry, and so on (ibid., 60, 49, 46). At the same time, however, the text suggests that the narratee is actually ignorant of QT. Thus chapter 2 (pp. 34–178) is announced as "a presentation of mathematical tools . . . i.e. a theory of Hilbert spaces and the so-called Hermitean operators [and] an accurate introduction to unbounded operators" (ibid., viii). The leisurely pace of this chapter strengthens the impression of an introduction, as do various comments in which the narratee is provided with little exercises to work through "whose discussion can be carried out without difficulty" (ibid., 60). At the end of the book the impression is consolidated retrospectively by elevating the narratee to the status of one who can now address "more complicated examples" (ibid., 495).

This construction of the narratee can be understood as a response to a rhetorical difficulty. The text's criticisms of the physics community, which is its intended audience, suggests that physicists were, by and large, unsympathetic to von Neumann's project—the construction of QT upon proper mathematical foundations. Dirac's less demanding, nonrigorous approach was preferred. The text is unable to argue too vigorously against these adversaries, however, since to do so would violate the official ideology against dissent within science. Nevertheless the text needs to pursue the argument and convert its readers to von Neumann's cause since its authority depends upon the good opinion of the scientific community (communal accord confers authority in science). Here lies one of the roots of the text's rhetorical difficulties: How can it convince its skeptical readers of what they reject if they are the source of its credibility?

The text's response is a new version of QT built along suitably rigorous lines but presented in the narrative form of an education for the novice at QT. This narrative construction conceals the text's role as a radical critique and re-education of the already (mis)educated expert at QT who is the text's implied reader.[5] The narrative of pedagogy does not, however, fit altogether smoothly into the text; on the contrary, it is accompanied by certain structural flaws that repeat the text's basic ideological contradiction.

The preface declares, "The object of this book is to present the new QT in a unified representation which . . . is mathematically rigorous (ibid., vii). This foreshadows the book's historical narrative which covers the development of a "new" QT from the "old", but also anticipates a pedagogic narrative which introduces the reader to a version of the new QT constructed within a mathematically rigorous framework in which Planck, Einstein, and Bohr "dominate" the development of the old QT, 1900–1925; then Born, Heisenberg, Jordan, Dirac, and Shrödinger develop the new theory; and finally, Dirac takes a wrong turn from which the narrator of the text rescues the new QT—see ibid., 3; chap. 1, fn. 5; and 5.

The pedagogic narrative unfolds a temporal order embedded within the overarching historical narrative. The complex relation between these different temporal orders is apparent in the tense structure of chapter 1. Section 1 of that chapter is written predominantly in the past tense. It engages purely with the historical narrative, the unfolding in the recent past of the new QT, and ends with a present tense statement which brings the reader up to date with the state of QT, "[At] least there is a mechanics which is universally valid, where the quantum laws fit in a natural necessary manner, and which explains satisfactorily the majority of our experiments" (ibid., 6); indeed this is "the present state of affairs" (ibid., 6, fn. 10). Present tense, sustained in the first six pages of chapter 1, section 2, gives way to a new temporal ordering, corresponding to a shift from history to pedagogy, as in "Let us set forth briefly the basic structure of the Heisenberg-Born-Jordan "matrix mechanics . . ."" (ibid., 6). Here, for the first time, the narrator addresses the narratee directly, interpellating the narratee ambiguously within the scope of the narratorial "we," and thus arranging that narratee will join narrator in a voyage of discovery of the true QT. The time within which this voyage unfolds is embedded within the extended present moment of the historical narrative and provides a framework within which the new QT is transformed into a more desirable, mathematically precise version. The horizon of events which define this transformation is constituted by the narrator's and narratee's engagement with the text. The narrator goes through a succession of explanations, and on this basis the narratee achieves a succession of new understandings, marked in the text by locutions such as "we must now learn as much as possible" (ibid., 7) and "it should be noted that" (p. 9), from which there are frequent analepses as in "which was mentioned above" (pp. 10–11), and prolepses as in "the operator calculus to be developed later" (p. 11, fn. 14). (Note, however, that the narratee's order of reading does not coincide with the implied reader's. The latter is not in a student relationship with the narrator, but rather in a collegiate one with the implied author. Therefore, the reading will be expected to break with the natural didactic order. Signs of this occur in the parenthetical notes to the reader, as in "the reader will find exhaustive treatments of this subject cited in Note 1" [1952, 9].)

But now consider the role which the text assigns to chapter 2 in which it introduces the mathematical preliminaries for a rigorous exposition of the new QT. At the beginning of chapter 3 in which the new QT is expounded in rigorous form, the text says, "Let us now return to the analysis of the quantum mechanical theories which was interrupted by the mathematical considerations of Chapter II" (ibid., 196). Here, chapter 2 is referred to retrospectively as an "interruption" to the overall project of the book (viz., the expounding of QT), one which is necessary in order to set the scene for the project's proper completion.

This explicit reference to chapter 2 as an interruption suggests that a continuity between chapter 1 (the history of QT) and chapter 3 exists, which in turn implies that the new QT expounded in the third chapter is simply a continuation of the old QT into the present. This continuity confers an authority upon the new QT of chapter 3 by presenting it as the natural inheritor of the successes of the old QT. The claim that chapter 2 is an interruption also de-emphasizes the text's criticisms of Dirac since the sole difference between the text's approach to QT and Dirac's lies in the rigorous foundations laid down in chapter 2. Thus labeling chapter 2 as a digression can be interpeted as a manifestation of the text's desire to suppress its radically critical thrust and identify more closely with the official discourse of science which emphasizes accord between colleagues and careful management of disagreement.

However, the text's explicit suggestion that chapter 2 is an interruption is contradicted by the text's narrative structure. Chapter 1 engages in an historical narrative, the development of QT, while chapter 3 introduces the rigorous reworking of present QT. Thus chapter 2 cannot be a mere digression since it falls between chapters which belong to different narratives. And since chapter 2 comprises nearly one-third of the book, additional doubt is cast upon the claim that it is a mere interruption. This contradiction between the narrative structure which foregrounds chapter 2 and the explicit self-misrepresentation of the chapter as a digression repeats the basic ideological contradiction between communal and individualistic conceptions of science.

We have seen that the *Gründlagen* repeats fragments of two rival ideological formations—an ideology of communalism internal to science and an individualistic ideology external to it—so that the basic ideological contradiction recurs at various levels within the text as dissonances within the narrative voice, narrative structure, and the alternately critical and conciliatory attitudes the text takes to its author's rival, Dirac. Each of these recurrences helps to minimize or conceal the basic ideological contradiction upon which it comprises a variation.

For example, the alternation of plural and singular referents for the narratorial voice repeats the ideological split between Mertonian and individualistic conceptions of science, but this split is concealed behind the uniform or neutral surface appearance of the narration, which is either a first-person plural "we" or impersonal (passive or transactional) narrative style throughout. In short, the text's surface rhetoric manifested in its grammatical forms helps to conceal the contradictory relations between the ideological fragments which impact upon the text. This process of concealment also enhances the appearance of a coherent narratorial position upon which the text's authority rests.

Desire in the Text

The *Gründlagen* reflects a distorted and fragmentary image of the ideological field in which it was produced. But texts have an ideological role which goes beyond mere distortion of independently constituted ideologies; they may also be sites at which ideology is constituted via the creation of desire.[6] The key to understanding texts' constitutive role in this respect is Freud's notion that from their very first moments subjects are haunted by the experience of their own lack, an experience which they are constrained to repeat in endless variation for the rest of their lives.

This psychic experience of self-lack resembles the social experience of ideological contradiction in a crucial respect: Both experiences bear down upon subjects, threatening the integrity of their relation to themselves and their surroundings. In virtue of this structural resemblance, subjects' experiences of ideological contradictions become means of repeating their experience of self-lack, and in this way the social difficulties associated with lived ideological contradictions assume the private meanings of loss and anxiety associated with subjects' sense of self-lack. The social thus becomes an arena in which subjects may dwell upon and work over their own defective relations with themselves.

According to Freud, subjects respond to endlessly recurring experiences of self-lack with games of desire in which by settling upon and aiming toward something which they want, subjects establish some measure of integrity. Lacan argues that such identification of desire always proceeds in two steps: one, an identification of what the Other desires; and two, an identification of the subject's own desire in relation to the desire of the Other.[7] In the special context of reader engagement with the text, the role of the Other is played by the voice of ideology mediated by the text.[8]

The relation of the *Gründlagen* to its readers is illuminated by these points. The text desires that its reader conform with a particular image of mastery, of mathematical rigor which it realizes in its practices and explicitly prescribes in its criticisms of Dirac. But the text's practices and its significations fail to uniquely determine the nature of that mastery. Moreover, by indicating that Dirac and the rest of the scientific community decline the rigor it exemplifies, the text indirectly raises the question of whether such rigor is really desirable, whether it is von Neumann rather than Dirac who is unconventional.

The *Gründlagen* raises but fails to resolve these questions and thus sets in motion significations which contradict its image of mastery. This contradiction repeats the text's fundamental ideological tension, which in turn is a means of repeating the basic incoherence in the reader-subject. In response, readers are led to establish their own desires as well as those of the Other who stands behind the voice of the text's incoherent ideological commands. More specifically, through their individual practices, readers demonstrate what they take the Other to want and what they propose to do about it, thereby constructing their own desires in relation to the desire of the Other.[9] For instance, some readers will take the text to be advocating a purely formal approach to scientific theorizing which they then reject in favor of less formal procedures, while other readers will see the text as allowing a role for metaphysical speculation. In short, the text leads readers to create their desires instead of merely appealing to (or persuading) them in terms of already constituted desires. Thus, as Lacan argues in connection with visual representations, there is no question of the text entrapping or seducing readers by promising them something they already desire, "This relation is not, as it might at first sight seem, that of being a trap for the gaze" (1981, 101). Instead, reading the text captivates readers by inciting them to construct and engage in the pursuit of a desire that distracts them from what they really want and know they cannot get. (For a general discussion of these points, see "Of the Gaze" [ibid., esp. 82–103, 109–116].)

Further exploration of these issues takes us beyond what we can discuss here. Suffice it to say that the text's dissonances are sites at which both rhetoric and ideology are engaged. They are places at which the text shapes readers' desire, but also sites at which rhetoric in a more traditional sense performs its task of concealment, a task that coincides with the traditional ideological one of misrepresentation. These dissonances are also sites of ideology in the Althusserian sense because, by acting as cradles for desire, they partake in the fundamental processes by which the imaginary relation which subjects inhabit are constituted.

NOTES

1. The English translation rather than the German original is used here in part because the translator's preface is one of the key sites at which the ideology of the text surfaces. The controversial issue of whether the stylistic and structural features of the translation are original to or part of the intentions of the author does not affect my conclusions since the text as a scientific production rather than as an element in a particular authorial oeuvre is my concern. The translation is referred to simply as "the *Gründlagen*."

2. For Althusser, ideology refers to the fundamental processes of constitution of the human subject but also refers to certain lived representations which take on an interpellative role. According to Althusser, these two approaches converge since subjects are created via a process of interpellation. Althusser also argues that ideological representations are imaginary in the sense that they are representations of the imaginary relations of subjects to their real conditions of existence. In the context of science such imaginary relations are constituted by scientists' loyalty to an institution which exploits their labor in the name of science and structures their practices in terms of the Baconian-Mertonian norms. See Althusser (1971, 152–65) for a general discussion of Althusser's conception of ideology.

3. On other occasions the narrator disappears entirely in what Fowler (1981) calls "generic claims," purely transactional generalizations, such as, "If R and S have inverses then RS has too." Although, more commonly such generalizations are presented in a passive form, such as, "If R and S can be inverted then RS can be inverted" or even as, "If R and S can be shown to have inverses then RS can be shown to have an inverse." These alternate forms are well illustrated in von Neumann (1952, 45). The latter variants upon the narrative voice efface the narrator and thus merely serve to focus attention upon the first-person plural form as the form of engagement with the narrator.

4. An opposition is being proposed here between an epistemology which evaluates scientific theories in terms of objective criteria of comprehensiveness and rigor, and an aesthetic approach to science, associated with Dirac, which privileges the subjective virtues of beauty and elegant sufficiency. The opposition can be seen as one between "romantic" and "enlightenment" approaches to knowledge. Kant's pejorative term *Schwärmerei*, meaning "buzzing"—as in the buzzing of bees, noisy, or intense but without substantive content—captures the manner

in which the romantics were figured by the enlightenment. (For a discussion see the introduction to Simpson 1984.) Eagleton (1982, chap. 4, 102ff.) discusses this conflict between enlightenment and romantic ideologies. In Eagleton's terms, the text being considered here must be seen as Brechtian in its rejection of a "nostalgic organicism" (ibid.), and in its allegiance to "open, multiple forms which bear in their torsions the very imprint of the contradictions they lay bare" (ibid.). However, simultaneously, the text must be seen as "organic" in its "traditionalist preference for closed symmetrical totalities" (ibid.). Here, as elsewhere, the text is in tension. This tension is echoed in von Neumann's other works in which scientific practice is presented as seeking for beautiful theories in order to entrap them within a suitably rigorous framework (1963, vol. 6, 477).

5. Note that the implied reader differs from the narratee, indeed, the text takes advantage of this difference to avoid an outright confrontation with its audience. Note too that the term "reader position" rather than "implied reader" is preferable here. Traditionally the latter is seen as an idealized representation offered by the text of its own readers, whereas the former is a position in terms of which a text interpellates its readers and thus one which is fleshed out by individuals' reading practices. It is therefore best seen as something the text displays or performs rather than merely represents, although it is also a representation of the reader by the text in the sense that it is cued by what the text says or implicitly takes for granted about the reader. See Chatman (1978) for a discussion of the traditional concept of implied reader.

6. The creation of desire implies the creation of imaginary relations of desiring. This additional dimension corresponds to the shift in Althusser's conception of ideology from the classical view, associated with the early Marx, that ideology is an imaginary representation of subjects' real conditions of existence to the view of the mature Marx that ideology is the medium within which subjects live (and represent) their imaginary relations to their real conditions of existence (see note 2).

7. Subjects do not always desire what the Other desires but their desires are always defined in relation to the desire of the Other. This is one meaning of Lacan's brilliantly equivocal claim that desire is always desire of the Other—de l'Autre. Its other meaning is that desire always involves a desire for the Other, a reference to the fact that the process of constituting desire is always fundamentally destabilized by an identification of the subject with the Other, or more correctly, with the position of the Other which is the position from which the subject appears desirable. In other words, in assuming a desire I not only project a certain idealized image of myself (what Freud calls the ideal ego) but also identify with a point of view (Freud's ego ideal) from which I appear desirable because I conform with that image.

8. Lacan defines the Other as the functionary which subjects construct as the point of address for their demand to get what they can never have, viz., their own sense of completeness. Because the voice of ideology repeats the primal lack of the subject, it will take on this role of the Other; that is, because the ideological command makes the subject aware of its own lack it creates a response in the subject

"What do you want" where the "you" is the construct which the subject erects as a point onto which it can project its own concerns. Lacan discusses the relation of text to Other in connection with the picture—see Lacan (1981, 105–22). For a revealing summary of Lacan's views on these points, sees Zizek (1989, chap. 3).

9. In practice, separating the two parts of this display may be difficult. For example, it may be impossible to decide whether my sloppiness is a competent response to a rejection of rigor or simply due to incompetence. The difficulty may not be merely epistemic but may reflect an ontological indeterminacy due to an underdetermination of desire by practice.

REFERENCES

Althusser, L. 1971. *Lenin and Philosophy and Other Essays*. London: New Left Books.
Chatman, S. 1978. *Story and Discourse*. Ithaca: Cornell University Press.
Eagleton, T. 1982. *Criticism and Ideology*. London: Verso.
Foucault, M. 1984. *The History of Sexuality*. Vol. 1, *An Introduction*. Harmondsworth: Penguin.
Fowler, R. 1981. *Literature as Social Discourse*. Bloomington: Indiana University Press.
Kress, G, and R. Hodge. 1979. *Language as Ideology*. London: Routledge & Kegan Paul.
Lacan, J. 1981. *The Four Fundamental Concepts of Psychoanalysis*. New York: Norton.
Macherey, P. 1978. *A Theory of Literary Production*. London: Routledge & Kegan Paul.
Merton, R. 1968. *Social Theory and Social Structure*. New York: The Free Press.
Simpson, D. 1984. *German Aesthetic and Literary Criticism: Kant, Fichte, Schelling, Schopenhauer, Hegel*. Cambridge, England: Cambridge University Press.
von Neumann, J. 1952. *Mathematical Foundations of Quantum Mechanics*. Princeton: Princeton University Press.
———. 1963. *Collected Works*. New York: MacMillan.
Zizek, S. 1989. *The Sublime Object of Ideology*. London: Verso.

12

Eddington and the Idiom of Modernism

Gillian Beer
Faculty of English, University of Cambridge

> All was fading, waves and particles, there could be no things but nameless things, no names but thingless names.
> —Beckett, *Molloy*

> In the scientific world the conception of substance is wholly lacking, and that which most nearly replaces it, viz. electric charge, is not exalted as star-performer above the other entities of physics. For this reason the scientific world often shocks us by its appearance of unreality. It offers nothing to satisfy our demand for the concrete. How should it when we cannot formulate that demand? I tried to formulate it; but nothing happened save a tightening of the fingers.
> —**Eddington**, *The Nature of the Physical World*

Arthur Stanley Eddington was not only one of the most distinguished astrophysicists of the 1920s and 1930s and one of the first to recognize the importance of Einstein's work, he was also, par excellence, the scientist whose writing attracted philosophers and the general public alike. His scientific reputation had been made as early as 1914 by the time of the publication of *Stellar Movements and the Structure of the Universe*. In 1919 he led the expedition to Brazil to observe the solar eclipse of May 29 which made it possible to verify Einstein's predictions. Writing in 1933 Lord Samuel remarked that Eddington's books "have attained a vast circulation" (p. 471) and that the debate aroused by his work on causation, relativity, and indeterminism "is no remote or unimportant discussion. It touches the very springs of thought and action in contemporary life" (ibid.). The extent of his popularity is indicated by *Science and the Unseen World* (1929), the Swarthmore lecture usually quietly received within the Quaker community. Within a year of its publication more than 20,000 copies had been sold.

In this essay I explore his more popular writing, particularly *The Nature of the Physical World* (1928), *New Pathways in Science* (1935), and *The Philosophy of Physical Science* (1939). All were based on lecture series, some were broadcast. These works—lucid, humorous, and innovative—had (and have) extraordinary appeal. They also raise a number of questions about the nature of authority, the constituencies of readers, and the interaction of discourses across diverse fields. They present a nicely insoluble conundrum about influence—a conundrum that makes it clear that we are likely to go astray if we look only for orderly reading exchanges between scientists and literary writers. It becomes clear that fugitive acquaintance, a sense of the possibilities opened by another's work, sometimes a single sentence on the first page ("The atom is as porous as the solar system"), may suffice to produce a charge that changes writing.

Perhaps most compelling for us now are the linked questions: How and to what extent does individual experience go to compose theory? and can scientific thinking give a purchase on communal ideology as well as being informed by it? I cannot compose full answers to these questions within the confines of this essay but Eddington is a suggestive case for scrutiny. His cultural influence was at its height in the 1930s. He was one of the first to make use of radio to broadcast complex ideas to a large and unknown audience. He was a birthright and convinced Quaker and thus committed to a number of causes that set him at jar with the establishment as well as being free from any imaginative commitment to *creed*. He is intriguing, too, because he was both politically aware and cloistered, a voracious reader of classic and popular literature, a conscientious objector who saw by 1920 the implications of the atom for peace and war. Unmarried, he was a fellow of Trinity College, Cambridge, and also kept house with his mother and sister. His father died before he was two, still *infans*, that is, without speech.

He was addicted to solving the crossword puzzles in *The Times* and the *New Statesman* (the voices of the establishment and of the dissident bourgeoisie). He usually finished within five minutes (Chandrasekham 1983). Crosswords are a repository of common knowledge in a particular class: They are complicit, knowing, and cipherlike. Yet in his own metaphorical systems within the books, Eddington is distrustful of ciphers, and he prefers game and drama, "Perhaps the *Tragedy of Hamlet* is not solely a device for concealing a cryptogram" (1935).

Much has been written in recent years about J. Derrida's (1970) use of the concept *jeu* (playfulness, gaming, the game with its inhering rules, the closeness in French between the sounds of first person and game, *je* and *jeu*). That concept has been translated into English in the form "free play" with its misleading emphasis on the random. In current discussion of metaphors for scientific inquiry the code (or cipher) and the game are often collapsed together. But in the cracking of a code, *resolution* alone is sought and *homology* achieved. In game, and in a play, on the other hand, a sequence of meaningful alternative and unpursued pathways provide a persisting complexity of choice reaching out from the chosen move. Other possibilities are always lying latent. For physicists in the 1930s, preoccupied with probability and indeterminacy, *game* or play became a more useful thought tool than coding and decoding.

One of Eddington's favorite texts of recourse is *Through the Looking Glass* (1872), a narrative based on chess moves and embracing a kind of nursery violence somewhat to Eddington's taste. That violence was expressed also as an eclectic ransacking of varied linguistic registers—yet it was always held nicely in play. The reviewer in *The Nation*, C. Singer, remarked on the index of *The Nature of the Physical World*, "It is not perhaps altogether surprising that the entry 'Hamiltonian differentiation' should be followed by Heaven, but it would have taken a clever person to guess that 'Solar system, origin of' should immediately follow 'Slithy toves'" (1928, 469–70).

E. Schrödinger (1935) argued that the "surplus force" (p. 23) beyond that needed to preserve life "manifests itself in play" (ibid.), within which he included sport, board games, and "research work proper" (ibid.). J. Jeans provided another important sense of "play" as loose jointedness, allowing a greater range of positions, "the fact that loose jointedness, of any type whatever, pervades the whole universe destroys the case for absolutely strict causation, this latter being the characteristic of perfectly-fitting machinery" (1930, 27–8). Eddington emphasized the making of stories and the activity of drama as ways of thinking about a world of physics that eschews models, machines, pictures, and substance itself. The idea of the drama and of functional roles particularly appeals to him because it allows for multiplicity of subject positions as well as demanding communal (and scripted) interaction.

What kind of story did Eddington enjoy making? Since his schooldays he liked to make tales that flirt with undecidability and that proffer and whisk away happy endings. In one of his schoolboy essays, now in the Wren Library, Trinity College, Cambridge, he describes "A Total Eclipse of the Sun." He foregrounds enigma: To learn about the sun it is necessary that it be not seen:

Except for the sunspots almost everything we know of the physical condition of the sun is derived from total eclipses. It seems paradoxical, but still it is true that by far the most things have been discovered about him from observations made when he was invisible. But for these short periods of invisibility . . . we should never have known what a marvelous body the Sun is. And truly he is a marvelous body. (Eddington Papers [EP] 0.11 22 1, v6–r7)

(The schoolboy Eddington always genders the sun masculine with upper case and the earth and moon feminine, with lower case. He grew out of it.) This paean of praise is followed by a lambent—and technical—description of the event. Yet he also observes the impropriety of adopting a single authoritative position for interpretation:

The term 'Eclipse of the Sun' is really not correct. In any other case the eclipse of a body means that it is obscured by the shadow of another body. . . . It is interesting to notice what our eclipses and other configurations of bodies would seem like when seen from other objects. Thus a total eclipse of the Sun (or as it should really be called an "occultation" of the Sun), from the moon would seem to be an annular eclipse of the earth, and from the Sun it would be called a transit of the moon. An annular eclipse of the Sun (which is really a transit of the moon) would be also a transit when seen from the Sun. From the moon it would be called an opposition of the earth. Again a total eclipse of the moon would seem a total eclipse or rather occultation of the Sun and be visible from quite half of the whole surface of the moon. From the sun it would be an occultation of the moon. (Ibid., 2r–2v)

He enjoys the knowing schoolboy tongue-twister, but there is more. His pleasure in the dance of terms here (occultation, transit, eclipse); the repositioning of language according to the place of observation; the fictional subject positions; and the use of humor and deixis already at the age of 16 in 1898 propose a relativized intellectual universe.

When he revised the manuscript of *The Nature of the Physical World* Eddington again and again recognized the need to emphasize *representation, attribution,* and the *activity of interpretation* rather than to affirm the essential properties of things. So, after revision, "fields of force" no longer *"belong"* but *"are assigned to"* "the category of *'influences'* not of *'things.'*" He draws progressively further away from the taken-for-granted categories by which classification takes place. "We feel it necessary to [*have*] [*assume*] concede some background to the measures" and—later in the same sentence—"the *properties* of this world" become "the *attributes* of this world" (EP, MS 2, xiii).

In each case the pressure of the statement shifts toward a recognition of human agency (such as assigning, conceding, attributing) and away from a substantiating prior condition to which reference may securely be made. That is, his theoretical stance demands that he foreground human activity of mind in his description. So this is not a matter simply of cajoling an amateur audience with personifications and invocations that will draw the audience persuasively into the unfamiliar discourse of physics. It is, rather, a purposive, hard-working attempt by a physicist to establish an epistemology that will not reinstate the absolute, the out-there, in its language even as it attempts to destabilize them in its argument.

He reinforced this emphasis on the human activity of representation and its inevitably multiple nature in *New Pathways*. The working physicist "had stumbled upon a multiplicity of representation" (Eddington 1935, 19) and "was very much bothered by it" (ibid.). Einstein's theory of relativity is, he argues, "such an epoch-making breach with tradition" (ibid.) because it necessitates "fluidity of representation" (ibid.) and abolishes the old kind of storytelling with its one authoritative message—the kind associated by novelist Virginia Woolf with determined surface enumeration. This process of destabilization gives Eddington problems when he wants to refer to matters beyond physics, as in the much rewritten sentence in *The Nature of the Physical World* that first reads "the phenomenon of contact of the external world of physics with human consciousness is outside the scope of physics." He revises it away from phenomena toward process and transformation. The potentially misleading word "matter" is changed to the vaguer "world," "becomes" is intensified to "is transformed into," and—perhaps most tellingly—"familiar acquaintance *with* human consciousness" becomes "familiar acquaintance *in* human consciousness." The sentence at last reads, "But the process by which the external world of physics is transformed into a world of familiar acquaintance in human consciousness is outside the scope of physics" (1928, xiv).

These detailed instances demonstrate the attentive microlevel of semantics at which Eddington was attempting to specify the modern world of physics. They demonstrate too that he is sometimes driven to choose the *slacker* word in order precisely to *tighten* his position.

Popper offers an account of how conceptual systems operate within language that helps to place some of the communicative problems that Eddington confronted. It is also unexpectedly close to the arguments of deconstructionist critics such as Derrida:

Any attempt to make the meaning of a conceptual system "precise" by way of definitions must lead to an infinite regress, and to merely *apparent* precision, which is the worst form of imprecision because it is the most deceptive form. (This holds even for pure mathematics.) (Popper 1982, 44)[1]

Eddington rummages through the ordinary world of his readers and listeners for examples in his argument, yet he feels the gap between what physics can say and the domestic epistomologies of daily life. He reaches for politics and finds it hard to place them. His writing places *him* in the midst of the literary modernists, some of whom—including Woolf and Empson—allude directly to him in their works. Yet seemingly in all his capacious reading he did not read Joyce, Woolf, Pound, or Eliot (the latter the most surprising). He did read Sinclair and some Lawrence (including, when it first appeared, what he calls with an intriguing swerve toward his own name, Eddington, *Lady Chatterton's Lover [Lady Chatterley's Lover]*).

The enormous reading lists that lie hidden in his 1905–1914 journal (augmented and brought up to date at the end of the 1930s) record when and how often he read Dante and Dorothy Sayers, Bertrand Russell and Burns, Poincaré and Poe, Agatha Christie and "In Memoriam", Emerson, Lewis Carroll and the *Iliad*, and Zola and P.G. Wodehouse. At the same time—both a practical difficulty and an opportunity for my argument—the lists are not all-inclusive. For example, he cites Bergson in *The Nature of the Physical World* but not in his lists. So (to take a minor example) he may or may not be echoing and reworking Eliot's now famous line, "This is the way the world ends, not with a bang but a whimper" when he concludes a paper on "The Source of Stellar Energy" in *Nature* (1926) by emphasizing that rather than "leading up to some great climax" (p. 32) his argument must "fizzle out" (ibid.), "I do not apologize for the lameness of the conclusion, for it is not a conclusion. I wish I could feel confident that it is even a beginning (ibid.). Questions of science, reason, and rhetoric are not separable in the case of Eddington, but there are conscious stresses within them.

Eddington was troubled by "the way in which the scientific realm of thought has constituted itself out of a self-closed cyclic system" (1928, 351). Since the emphasis of his thought is on relativizing he is cha-grined by the autonomy-seeking model of scientific discourse. Indeed, he mocks the narrative procedures underlying physics by a nursery analogy, "The definitions of physics proceed according to the method immortalized in 'The House that Jack Built'. . . . But instead of finish-ing with Jack, whom of course every youngster must know without the need for an introduction, we make a circuit back to the beginning of the rhyme. . . . Now we can go round and round for ever" (ibid., 262). He sees classical physics as foisting "a deterministic scheme on us by a trick, it smuggles the unknown future into the present" (ibid., 308) and thus produces what Eddington calls "retrospective symbols" (ibid.) that control our predictions.

Passages such as these allow W. Empson in his poem "Doctrinal Point" to pinpoint Eddington's "insolence"—a positive quality for Empson, allying him with the poet who refuses the containing circle of determined explanation:

> Professor Eddington with the same insolence
> Called all physics one tautology;
> If you describe things with the right tensors
> All law becomes the fact that they can be described with them;
> This is the Assumption of the description. (1955, 39)

So it becomes *conceptually* necessary for Eddington to reach be-yond physics for his discussion of physics. His evidences and his order-ings tend to take the form not of models or pictures but of utterances, dramas, and narratives.

Instead of the privileged anonymity of the single scientific ob-server—the passive voice, the "one" of discourse—he implicates him-self and reader, foregrounding the I-you relation whose relations are never fixed. The observer is in the experiment, the reader in the text. The reader is given a voice, interrupting, questioning, not always satisfied by Eddington's responses. This move is not persuasive only; it is an enactment of the theory he expounds. Observation and spacetime are relativized. *The Nature of the Physical World* begins, "I have settled down to the task of writing these lectures and have drawn up my chairs to my two tables. Two tables! Yes, there are duplicates of every object about me—two tables, two chairs, two pens" (1928, xi). *New Pathways in Science* opens with the sentence, "As a conscious being I am involved in a story" (1935, 2). The second paragraph opens, "As a scientist I have become mistrustful of this story" (ibid.). The third paragraph complicates the relations between these two positions of involvement and skeptical construal. He must avoid "overmuch confidence in the storyteller who lives in my mind" (ibid.) but ignores him at his peril, "For I am given a part in the story, and if I do not take my cue with the other actors it is the worse for me" (ibid.):

For example, there suddenly enters into the story a motor car coming rapidly towards the actor identified with myself. As a scientist I cavil at many of the particulars given by the story teller—the substantiality, the colour, the rapidly increasing size of the object approaching—but I accept his suggestion that it is wisest to jump out of the way. (Ibid.)

Here we have the inverse of the efficacy of science argument: the inefficacy of scientific description in warding off the blow. It is necessary for human survival, he insists, to live with multiple epistemologies. Eddington favors stories with death in them—and the comedy of brief survival. He emphasizes this rhetorically in the following passage by setting the genre, "epic," against meager "inadvertence":

Nature seems to have been intent on a vast evolution of fiery worlds, an epic of milliards of years. As for Man—it seems unfair to be always raking up against Nature her one little inadvertence. By a trifling hitch of machinery—not of any serious consequence in the development of the universe—some lumps of matter of the wrong size have occasionally been formed. These lack the purifying protection of intense heat or the equally efficacious absolute cold of space. Man is one of the gruesome results of this occasional failure of antiseptic precautions. (Ibid., 309)

We become bungle instead of aim. The slang "gruesome" is gothicized by the suggestion of infection in the subsequent "antiseptic." He humbles the human assumption of our own centrality while at the same time enjoying the hubristic authority of being the storyteller. The reader is allowed to share that glee in a way that restores the centrality of mind to his explanation even as he dashes our pride. And that is one of his subtlest maneuvers, for the contentious conclusion he insists on reaching is (in a careful, cumbersome formulation uncharacteristic of his spry style) "that the world-stuff behind the pointer readings is of nature continuous with the mind" (ibid., 331).

J. Huxley (1933) argued that each age has its own antinomies, the current ones being science and human nature. Leonard Woolf (1929) asserted that "the spectacle of the physical universe from every human point of view, is horrible and terrible in its meaninglessness. . . . And it seems to me that the free human spirit should leave it at that and not try to console itself with either the old or the new fairy tales" (L. Woolf 1929, 829). Eddington is not willing to accept those antinomies, but neither can he knit them up. In *Science and the Unseen World* (1929) he asserts that "Human personalities are not measurable by symbols any more than you can extract the square root of a sonnet" (p. 33). But the relations of physics and human personality had changed in his lifetime—and so had the meaning of "physics" and of "human nature." Virginia Woolf had argued that about 1910 human nature changed: This was her hyperbolic way of releasing fiction from all its older conventions of representation. In her work words like the "real" and the "substantial" come under stress, just as much as they do in the work of Eddington. Her most experimental work is *The Waves* (1931). As she wrote it, she was reading Jeans's *The Universe about Us* (1929) and was trying to imagine space bending backwards. Eddington's work was frequently discussed in *The Nation*, of which her husband was literary editor and she read his work, though when is unclear. Yet her novel has been traditionally read as if it had nothing to do with a world in which everything had resolved into waves.

Virginia Woolf, like most educated people of the 1920s, was well aware of Einstein as an intellectual presence. And, like many others, she found his theories both baffling and magical. On 20 March 1926 she went to a supper party, "I wanted, like a child, to stay and argue. True, the argument was passing my limits—how, if Einstein is true, we shall be able to foretell our own lives" (1980, 68). That dazzled appreciation (or misprision) of Einstein's work haunts the writing of the 1930s. In the main, her understanding came through newspaper articles and through the writing of Eddington and Jeans (though we should note that she also pasted in her scrapbook Einstein's denunciation of fascism). However, any too conscientious complete approximation of Woolf to the principles of relativity would have to cope with her amusement at any such idea. In May 1938, the first person to attempt to write a dissertation on Woolf visited her, "Miss Nielsen came: a donnish bee haunted American lit prof., entirely distracted by Einstein, and his extra mundane influence upon fiction. L. threaded the maze to the muddle in the centre. I gave up on the outskirts" (1984, 146).

Still Nielsen perhaps had her effect. Woolf resists particularly her wholesale claim, but Einstein and Eddington are scattered in among other ideas in the gossip of *Between the Acts* (1941) and Eddington and Jeans are named alongside Darwin there. Toward the end of the book, when the play has ended and the audience is mulling over the afternoon, the hubbub of conversation yields vagrant talk of oracles and crepe soles, physics, religion, the onset of war, red fingernails, and sex. It includes this passage:

Oracles? You're referring to the Greeks? Were the oracles, if I'm not being irreverent, a foretaste of our own religion? Which is what? . . . Crepe soles? That's so sensible . . . They last much longer and protect the feet. . . . But I was saying: can the Christian faith adapt itself? In times like these . . . At Larting no one goes to church . . . There's the dogs, there's the pictures. It's odd that science, so they tell me, is making things (so to speak) more spiritual . . . The very latest notion, so I'm told is, nothing's solid . . . There, you can get a glimpse of the church through the trees. . . . (Ibid., 232)

From the 1920s on at least, Woolf found in the new physics dizzying confirmation of her sense that the real and the substantial are not the same. As early as 1923 she had invented a word for what she wanted to do, "to 'insubstantise.' " Eddington in 1928 describes a similar project, "[T]he solid substance of things is another illusion. It too is a fancy projected by the mind into the external world" (p. 318). When Woolf comments wryly of *The Waves* (1931) that there is "no quite solid table on which to put it" (p. 46), and when within the novel her thinkers doubt "the fixity of tables" (p. 315), "are you hard?" (ibid.), she is taking part in a stressful debate at large in the world of her time, not engaging in solipsistic prose poem musings on her own. Eddington suggests that substance is the hardest comfort to do without, "[We] have seen that substance is one of our greatest illusions . . . . Perhaps, indeed, reality is a child which cannot survive without its nurse illusion" (1928, 318). And in a virtuoso passage at the end of the chapter "Science and Mysticism" (Woolf wrote of *The Waves* also as "an eyeless, mystical, book"), Eddington describes the difficulties that the scientist experiences on the threshold of a room in terms like those that Rhoda experiences in the novel as she approaches the puddle she cannot cross: For each of them the solid world dislimns. The improbability of physical life declares itself. Typically, it combines a story which is about the difficulty of beginning a story, bodily buffetting, insubstantiated and yet a vigorous bodily presence, and the parabolic suggestion that those rich in knowledge are poor in advent. Even the eye of a needle presents no greater difficulties for entry than the doorway to a room:

I am standing on the threshold about to enter a room. It is a complicated business. In the first place I must shove against an atmosphere pressing with a force of fourteen pounds on every square inch of my body. I must make sure of landing on a plank travelling at twenty miles a second round the sun—a fraction of a second too early or too late, the plank would be miles away. I must do this whilst hanging from a round planet head outward into space and with a wind of aether blowing at no one knows how many miles a second through every interstice of my body. The plank has no solidity of substance. To step on it is like stepping on a swarm of flies. Shall I not slip through? No, if I make the venture one of the flies hits me and gives a boost up again; I fall again and am knocked upwards by another fly; and so on. I may hope that the net result will be that I remain about steady; but if unfortunately I should slip through the floor or be boosted too violently up to the ceiling, the occurrence would be, not a violation of the laws of Nature, but a rare coincidence. These are some of the minor difficulties. I ought really to look at the problem four-dimensionally as concerning the intersection of my world-line with that of the plank. Then again it is necessary to determine in which direction the entropy of the world is increasing in order to make sure that my passage over the threshold is an entrance, not an exit.

Verily, it is easier for a camel to pass through the eye of a needle than for a scientific man to pass through a door. And whether the door be barn door or church door it might be wiser that he should consent to be an ordinary man and walk in rather than wait till all the difficulties involved in a really scientific ingress are resolved. (1935, 342)

Notably, W. Benjamin ([1938] 1973) singles out and quotes this same passage of Eddington's in his second essay on Kafka, claiming that "If one reads [it,] one can virtually hear Kafka speak" (p. 145). He may also have been struck by the likeness between the topic of the threshold in Eddington—here and in Kafka's massive parable "Before the Door" in *The Trial*. Benjamin comments on the tensions of "physical perplexity" and "mysticism" in Eddington and Kafka and swings his argument round into a blast of terrible foresight which includes his own individual death as part of "the masses . . . being done away with" (ibid., 145):

that reality of ours which realizes itself theoretically, for example, in modern physics, and practically in the technology of modern warfare. What I mean to say is that this reality can virtually no longer be experienced by an individual, and that Kafka's world, frequently of such playfulness and interlaced with angels, is the exact complement of his era which is preparing to do away with the inhabitants of this planet on a considerable scale. (Ibid., 145–46)

Benjamin pinpoints ways in which the nondeterministic, relativized, ethereal world of play and representation in physics, and in its accounts of itself in the 1930s, are, ultimately, terribly yoked to the destructive outcomes of those same physics. The personal politics of Eddington are those of the conscientious objector, of the peace movement and Quaker humanitarianism. Eddington (1935) ends his discussion of "subatomic energy" thus:

It cannot be denied that for a society which has to create scarcity to save its members from starvation, to whom abundance spells disaster, and to whom unlimited energy means unlimited power for war and destruction there is an ominous cloud in the distance though at present it be no bigger than a man's hand. (P. 163)

He lives to find that the work of physicists is fundamental to the conduct of the war, culminating in the atom bomb at its end. The politics of Virginia Woolf are similarly antimilitarist, Labour. Increasingly as the 1930s goes on she clings to the frail hope of the Peace Pledge movement and grits her teeth in the knowledge of the coming war. In *Three Guineas* (1938) she seeringly brings together gender politics and the politics of current rearmament and masculine aggression. At the same time she found herself accused in writing by her nephew J. Bell of being part of the quietist artistic world that had not acted politically when it was needed. Now the war was inexorably approaching which they had done too little to prevent.

The common danger for artists and physicists at the time is that of mentalism, a withdrawal from the social fray—particularly the political fray, in the light of an emphasis on emptiness, porousness, atomism, undecidability, and "mind." Doing away with a stable nature "out there" where "there was no great difference between appearance and reality" (Jeans 1933, 2) and substituting a position in which the scientist (and the novelist) must see nature not as "entirely distinct from himself" (ibid.) rather, "Sometimes it is what he himself creates, or selects or abstracts; sometimes it is what he destroys" (ibid.). Fascist art did not preoccupy itself with insubstantiality. Heads were substantial until the fascists got to them. Einstein and others were well aware of the quietism that could be the outcome of the loss of purchase for the physical world that the new physics could imply.

Modern physics, Eddington argued later, is *"off the gold standard"* (1935, 81; emphasis in original). That is, it no longer has an unchanging point of reference or conversion. Instead of seeking always a return to a stable equivalence with a steady origin (gold), it must describe the shifting relations between currencies (and suddenly the meaning "flow" comes back into the word). Instead of a determined set of causalities, it must accept that "the only reason assigned for any regularity is that the contrary is too improbable" (ibid., 79) and "'Impossible' therefore disappears from our vocabulary" (ibid., 80). That absenting is peculiarly terrible in the work of Kafka, where repetition, formulaic encounters, and ritualized *wiederholungszwang* (in Freud's term) are all insufficient to ward off the appalling—and comic—recognition that the possible has no bounds.

Indeed, there is something Kafka-esque in Eddington's resubstantiating of the mutilated observer produced by Einstein's language:

[A] peculiar person called the observer—the man who has a habit of falling down lifts, or getting transported by aeroplanes travelling at 161,000 miles a second. . . . He has one eye (his only sense organ) which is colour-blind. He can distinguish only two shades of light and darkness so that the world to him is like a picture in black and white. The sensitive part of his retina is so limited that he can see in only one direction at a time. We allow him any number of assistants equipped like himself so that they can keep watch on the different parts of an experiment and pool their knowledge afterwards. Since we have so mutilated him he cannot make the experiments himself. We perform the experiments and let him keep watch. The point is that all our knowledge of the external world as it is conceived to-day in physics can be demonstrated to him. If we cannot convince *him* we have no right to assert it. (Ibid., 13)

This ghoulish comedy seems to harbor fears concerning the outcome of what he saw as the obeisance of physics to a lopped idea of the human:

Is it merely a well-meaning kind of nonsense for a physicist to affirm this necessity for an outlook beyond physics? It is worse nonsense to deny it. Or as that ardent relativist the Red Queen puts it, "You call that nonsense, but I've heard nonsense compared with which that would be as sensible as a dictionary" (1928, 344)

The rhetorical questions ward off the charge against him of being "well-meaning": seeking more sanguine stories than the universe freely provides, groping toward a multivocality that denies too little, and avoiding the astringent immediacy of the one meaning. Russell in *The Scientific Outlook* (1931) (itself a title with at least three meanings: prospect, future, and mindset) is tart about the "nice guy" aspect of then current physics, "The new philosophy of physics is humble and stammering, where the old physics was proud and dictatorial" (p. 88). Russell likes and admires Eddington but is skeptical about the temper of his earlier narratives. Eddington, Russell argues:

[p]roceeds to base optimistic and pleasant conclusions upon . . . scientific nescience . . . . [I]t is difficult to see what ground for cheerfulness modern physics provides. It tells us that the universe is running down, and if Eddington is right, it tells us practically nothing else, since all the rest is merely the rules of the game. (Ibid., 96)
   Nothing further will happen except that the universe will gradually swell. It speaks well for Sir Arthur's temperamental cheerfulness that he should find in this view a basis for cheerfulness. (Ibid., 97)

Russell's grounds for attack on that nescience of science urged in Eddington's writing are political as much as logical. Like Samuel a couple of years later, he hints that the loss of a securely causative system will undermine science as "the source of practically all change both for good and evil" (ibid., 97)—even, he implies, undermine social change itself. Science will lose its power to institute change because nonscientists will no longer accept the authority of its sequencing; such seems to be his line of argument. Machines will survive, but no longer fueled by theory. Then, however, he mischievously places himself averse to all those who seek unifying or totalizing systems. This swerve brings him back alongside Eddington—or, rather, beyond Eddington on a continuum with postmodernism, "I think the universe is all spots and jumps, without unity, without coherence or orderliness or any of the other properties that governesses love" (ibid., 98).

Eddington was stung by the suggestion that his work was in league with reactionaries, undermining social reform. He incorporated his response to Samuel's attack in *New Pathways* and the darkening parables of that book (like the mutilated observer quoted earlier) are imbued with fears that do not find direct expression. Eddington recognized, too, that the material means of production for his communication was part of a new social politics in which overlapping groups, often at odds with each other, must be sounded. The constituencies he addressed in his more popular books were his fellow scientists, educated nonscientists, readers of Penguin books, and the mass audience of radio. The first of those constituencies (fellow physicists) he tweaks by the tail with his suggestion that the autotelic system of physics is fated to misdescribe the world because it cannot work with those counterepistemologies of the everyday. It can neither say sufficient, nor little enough. Russell gave a pithy description of the linguistic dilemma:

[O]rdinary language is totally unsuited for expressing what physics really asserts, since the words of everyday life are not sufficiently abstract. Only mathematics and mathematical logic can say as little as the physicist means to say. (Ibid., 85)

Eddington is willing to risk giving what Russell scornfully calls "a cheerful impression of something imaginable and intelligible, which is much more pleasant and everyday than what he is trying to convey" (ibid., 85). Russell's emphasis on the untranslatability of physics is dour, even obscurantist. Eddington's enunciation emphasizes rather the incommensurability between symbol and subject: He often has recourse to images from radio as a familiar example of physics at work in the world:

The mind as a central receiving station reads the dots and dashes of incoming nerve-signals. By frequent repetition of their call-signals the various transmitting stations of the outside world become familiar. We begin to feel quite a homely acquaintance with 1LO and 5XX. But a broadcasting station is not like its call signal; there is no commensurability in their nature. So too the chairs and tables around us which broadcast to us incessantly those signals which affect our sight and touch. (1929, 23)

Eddington worries at the problem of the resistance of the universe to human reason throughout his career. In his later work, *The Philosophy of Physical Science*, he quotes and offers a critique of an earlier statement:

Eighteen years ago I was responsible for a remark which has often been quoted:

It is one thing for the human mind to extract from the phenomena of

> nature the laws which it has itself put into them; it may be a far harder
> thing to extract laws over which it has had no control. It is even possible
> that laws which have not their origin in the mind may be irrational, and
> we can never succeed in formulating them.

This seems to be coming true, though not in the way that then suggested itself.
(P. 179)

The only revision that he in fact offers is to substitute for "possibly
irrational behaviour," "undetermined behaviour." He glosses, "The
totality of mind-made law does not impose determinism" (ibid., 180).
He dislikes the phrase "the law of chance" since it converts into a set
of conditions "what is merely an absence of law in the usual sense of
the word" (ibid., 181). He positively relishes the shiftiness of concepts,
as in his curious pleasure in the word *hollow*:

> Like the symbolic world of physics, a wave is a conception which is hollow
> enough to hold almost anything: we can have waves of water, of air, of aether,
> and (in quantum theory) waves of probability. (1935, 320)

The body of water waves which for Helmholtz "awakens in me a peculiar kind of intellectual pleasure, because here is laid open before the
bodily eye what, in the case of the waves of the invisible atmospheric
ocean, can be rendered intelligible only to the eye of the understanding" (1890, 2: 598) for Eddington on the contrary is pleasurable precisely because it deliquesces into a range of probabilities.

Emptying becomes ideal as well as activity. Turning again to that
favorite philosophical object, the table, Eddington remarks:

> As for the external objects, remorselessly dissected by science, they are studied
> and measured, but they are never *known*. Our pursuit of them has led from
> solid matter to molecules, from molecules to sparsely scattered electric
> charges, from electric charges to waves of probability. Whither next? (1935,
> 322)

This acceptance of absence, indeterminacy, and the unknowable as
part of theory comes much more easily to Eddington emotionally than
it does to many of his fellow scientists. It sets him alongside, and in
the midst of, literary modernists and surrealists. But it derives, in all
probability, from a set of positions that are ethical and religious in
their foundations. That oddly archaic form of question "Whither
next?" gives a clue. The mixture of quietism and social awareness in
Eddington's language and work draws on his upbringing and continued membership of the Quakers. One of the tasks of Quakers is to be

a "seeker," but what is sought is placed beyond language. The Quakers are, as he proudly remarks in *Science and the Unseen World*, not bound to any creed and can therefore be free seekers. Energy of inquiry and enigma combine. As a young man at the annual Quaker camp (still held) in the Lake District he heard lectures by R. Jones on "Mysticism," W. Franks on "Job," and two he particularly admired by S. Rowntree on social questions. He continues, "The devotional meetings were good but there was no silence; far too many little speakers" (EP, b. 48, 56).

Eddington's imagination and receptivity was formed along paths that valued search and unknowing at once, that distrusted fixed images, and whose own presiding metaphor was that of the "inward light." This gave him a particular relationship to the problem of language and physics that Russell pithily described.

As a Quaker, Eddington was used to being frugal and he put that training to work intellectually. But unlike the early Quakers who eschewed stories as a form of lie, Eddington used that same distrust of the authority of representation to equalize diverse forms of description and explore epistemological issues, "[T]he physical universe is defined as the theme of a specified body of knowledge, just as Mr. Pickwick might be defined as the hero of a specified novel. A great advantage of this definition is that it does not prejudice the question whether the physical universe—or Mr. Pickwick—really exists" (1939, 3). Whereas Einstein was troubled by the idea of a "boshaft" god who dices with the universe and Jeans was self-comforted by the idea of a god who was a mathematician like himself, Eddington seems undisturbed by multiple representation and welcomes the relativity of description that must follow in the wake of Einstein's work.

Fundamental also must be an emphasis on motion, not stasis. Here a problem inhered in discourse as opposed to mathematical symbol. Eddington comments on the paucity of verb forms and argues that "the view that activity (expressed by verbs and gerunds) is of a few simple kinds and that variety resides in passivity (expressed by nouns) has purely linguistic origin. The paucity of verb forms is familiar to mathematicians as a difficulty of ordinary speech easily surmounted in their own symbolic language" (ibid., 213). One problem the physicist faces, he suggests, is that of translation from a symbolic system already imbued with subtle gradations of temporality and motion into a cruder speech system. Yet he is also distrustful of the resolution by observing physicists of everything into waves. He argues that substance can be expressed now only as form, and that as *form in motion*, "[T]he concept of substance has disappeared from fundamental physics; what we ultimately come down to is form. Waves! Waves!! Waves!!! Or for a change if we turn to relativity theory—curvature!" (ibid., 110). For Jeans (1933, 11) substance becomes *event*, for Eddington *form* (without the implication of stasis), while for other popular interpreters such as J. W. N. Sullivan (1933, 45) substance becomes *behavior*.

Eddington's written eloquence and lucidity was renowned. So was his humor. These qualities relied upon a certain insouciance in his relation to representation. He seems never to take it too solemnly. His metaphors do not attempt to drop on all fours. His (1928) opening description of the "two tables" draws its force from disparity, the difficulty of keeping two representations simultaneously in mind. In that, the game he plays mimics the wave-particle dilemma itself. Thus he trains the reader in how to read modern physics: trust themes not objects, form not substance, and incommensurable narratives not single stories. Everywhere underlying his argument is an emphasis on storied sequences. but these sequences are not like nineteenth-century narratives, nor even like the detective stories he much enjoyed reading. Abduction is only part of the pleasure. As prominent is dissolution, false endings, wayward connections, and simultaneities. When *Nature of the Physical World* was reviewed in *The Nation* in 1928 the abutting review on the same page was of E. Muir's *The Structure of the Novel* (1928), which was to prove an important pioneer work in the analysis of narrative forms. The reviewer quotes Muir on the issues of stasis and kinesis, time and space, "In the dramatic novel, space is more or less given, and the action is built up in Time. In the character novel it is Time that is assumed, and the action is a static pattern, continually redistributed and reshuffled in space" (Dec. 29, 1928, 470). Both books under review appeared while Virginia Woolf was working on two works that would reshuffle both time and space: *Orlando* and *The Waves* which include many sentences such as "I do not see this actual table, except in flashes. I see a wave breaking upon the limits of the word" (1976, 296)

Eddington's persistent revelation is that the idea of immobile substance and objects must be abandoned in physics. But he grows uneasy with the degree of selection that this implies for the observer. In his later work, *The Philosophy of Physical Science*, written at the time of the onset of the second world war, he worries at the expunging of so much common experience from a theoretical world that claims inclusiveness. Reason revolts, he suggests, when we consider what to do with the leftover world in consciousness: not only familiar tables and redness, he writes, but "the passage of time, a guilty conscience, the taste of sugar, being in love, toothache, amusement at a joke" (1939, 194). And, to emphasize his distrust of the observer's self-fulfilling functions and self-confirming conclusions, he emends a well-known story: Does experiment put the atoms or the radiation into "the state whose characteristics we measure" (ibid., 109)? Parable-wise he continues:

I shall call this Procrustean treatment. Procrustes, you will remember, stretched or chopped down his guests to fit the bed he had constructed. But perhaps you have not heard the rest of the story. He measured them up before they left the next morning, and wrote a learned paper "On the Uniformity of Stature of Travellers" for the anthropological Society of Attica. (Ibid.)

In this shy Quaker lies a knowledge of violence which he controls and deflects, except in his stories. A disquiet haunts him about the leftover questions raised by scientific inquiry and set aside. Those questions are the social, economic, metaphysical issues of the baleful 1930s in Europe. His writing rarely engages directly with such issues but it played into the anxieties of his radio listeners and his readers. It could do so because the rhetoric of his books is a paradoxical *substantiation* of his theoretical positions (bringing the body back in, refusing to void it from his language) and a critique of the claim of physics to autonomy of signification.

NOTE

I am grateful to the Master and Fellows of Trinity College for permission to quote from the Eddington Papers.

1. In the course of an attempt to distinguish between theory and conceptual systems Popper italicizes "*A theory is not a picture*" (1982, 5). Yet, as he acknowledges, scientists such as Bohr and Einstein had moved away from models yet had recourse to the idea of "pictures": the "particle picture," "the wave picture," "the

man in the lift." Jeans fronts *The New Background of Science* (1933) with contrasting photo-images of electrons *"behaving as"* (p. 2) "wave" and "particle."

REFERENCES

Benjamin, W. [1938] 1973. *Illuminations*. London: Fontana.

Carroll, L. 1872. *Through the Looking Glass*. London: John Murray.

Chandrasekham, S. 1983. *Eddington: The Most Distinguished Astrophysicist of His Time*. Cambridge, England: Cambridge University Press.

Derrida, J. 1970. "Structure, Sign, and Play." In R. Macksey and E. Donato, eds., *The Language of Criticism and the Sciences of Man*. Baltimore: The Johns Hopkins Press.

Eddington, A. S. 1914. *Stellar Movements and the Structure of the Universe*. London: MacMillan.

———. 1926. "The Source of Stellar Energy." *Nature*.

———. 1928. *The Nature of the Physical World*. Cambridge, England: Cambridge University Press.

———, 1929. *Science and the Unseen World*. London: Allen & Unwin.

———. 1935. *New Pathways in Science*. Cambridge, England: Cambridge University Press.

———. 1939. *The Philosophy of Physical Science*. Cambridge, England: Cambridge University Press.

Eddington Papers (EP). Wren Library, Trinity College, Cambridge.

Empson, W. 1955. *Collected Poems*. London: Chatto & Windus.

Huxley, J. 1933.

Jeans, J. 1929. *The Universe About Us*.

———. 1930. *The Mysterious Universe*. Cambridge, England: Cambridge University Press.

———. 1933. *The New Background of Science*. Cambridge, England: Cambridge University Press.

Maxwell, J. C. 1890. *The Scientific Papers*. 2 vols. Cambridge: Cambridge University Press.

Muir, E. 1928. *The Structure of the Novel*.

Popper, K. 1982. *Quantum Theory and the Schism in Physics*. London: Hutchinson.

Russell, B. 1931. *The Scientific Outlook*. London: Allen & Unwin.

Samuel, H. (Lord). 1933. "Cause, Effect, and Professor Eddington." *Nineteenth Century and After* 113: 469–547.

Schrödinger, E. 1935. *Science and Human Temperament*. London: Allen & Unwin.

Singer, C. 28 December 1928. Review of *The Nature of the Physical World*. *The Nation*, pp. 469–70.

Sullivan, J. 1933. *Limitations of Science*. London: Chatto & Windus.

Woolf, L. 28 September 1929. "Ad Astra?" Review of Heard and Jeans. *The Nation*, p. 829.

Woolf, V. 1931. *The Waves*. London: Hogarth Press.

———. 1938. *Three Guineas*. London: Hogarth Press.

———. 1941. *Between the Acts*. London: Hogarth Press.

———. 1976. *"The Waves": The Two Holograph Drafts*. Edited by J. W. Graham. London: Hogarth Press.

———. 1980. *The Diary of Virginia Woolf*, vol. 3. Edited by A. O. Bell. London: Hogarth Press.

———. 1984. *The Diary of Virginia Woolf*, vol. 5. Edited by A. O. Bell. London: Hogarth Press.

Comment: Standing on the Threshold

Trevor Melia
Department of Communication, University of Pittsburgh

G. Beer's essay is especially interesting because it brings together two hitherto widely separated disciplines—physics, perhaps the sternest of the sciences, and rhetoric, the harlot of the arts. This unlikely union is, however, also the occasion for a no less significant *re*union between two long estranged traditions in rhetoric—the belletristic, which has primarily been concerned with schemes and tropes and the adornment of language generally, and the tradition, deriving from Aristotle, which regards rhetoric as the "counterpart of dialectic" and as a close relative of political science. Literary critics situated in departments of English have normally been custodians of the belletristic tradition, rhetoricians operating in departments of communication, citing not only the work of Aristotle, but also of the sophists Gorgias and Protagoras, Isocrates, and Cicero have, like their predecessors, maintained an interest in the practical affairs of the polis. It is not true, as is sometimes alleged, that the latter never read literature or that the former are entirely innocent of politics. It is gratifying to me, working from the political side of the divide, to find so much with which to agree in Beer's virtuoso reading of Eddington's work. What follows is hence rather by way of supplement than dissent.

Eddington's self-appointed task in the three books focused upon in Beer's study is nothing less than to convey to the intelligent laypersons who were his readers the gist of the revolution in physics that had occurred since the turn of the century. He was not the only scientist to undertake that daunting task but he was probably the most successful.

This in spite of the fact that Eddington sought for pedagogical reasons simultaneously to undermine the very authority of science that is typically the mainstay of scientific rhetoric. The problem for Eddington was that what was taken to be common sense by the educated layperson was undergirded by the authority of no less a scientist than Newton. Eddington's rhetorical tactic was a provisional dismantling of the Newtonian universe in order to make room for the modifications that relativity and quantum theory had recently made necessary. As Beer shows, Eddington brought modernist (not to say postmodernist) predilections to his project. Hitchhiking on Beer's insight I suggest that Eddington shared not only intellectual affinities with the likes of Derrida and Benjamin but that his *rhetorical* situation was in many ways also similar.

Derrida's deconstructionism was threatening enough to the custodians of the orthodox analytic tradition in philosophy that they condemned it as "linguistic Mau Mau" and swore that it should never be taught at Oxbridge. Cambridge University, seeking perhaps to allay suspicions that it understood neither African politics nor contemporary developments in literary criticism, later awarded Derrida an honorary degree. Beer was, of course, a leader in the battle against the reactionaries who resisted this rectifying gesture. Resistance to Eddington's "selective subjectivism" was hardly as vigorous, although such philosophers as C. E. M. Joad, W. Stace, and even B. Russell expressed misgivings, and among scientists names as resonant as Plank, Einstein, and Rutherford, each architects of the physical world picture Eddington sought to depict, could be counted among his opponents. Members of the lay audience who were the principal target of Eddington's books and talks must have been even more startled than his scientific colleagues who after all were well aware of the recent challenges to the Newtonian universe. Eddington introduced his audience to those challenges by describing the trip "from solid matter to molecules to sparsely scattered electric charges, from electric charges to waves of probability" (1935, 323). And if his summary of quantum theory was terse, Eddington's description of the effects of relativity was even more so, "Modern physics is *off the gold standard*" (ibid., 81). In so saying Eddington illustrates the skillful use of metaphor that is perhaps the most obvious feature of his rhetoric generally. He problematizes the commonsense view of science by posing a metaphorical question, "Is the bunghole of a barrel part of the barrel? Think well before you answer, because the whole structure of theoretical physics is trembling in the balance" (1939, 119). Recognizing that success at prediction is what gives science its credibility, Eddington, invoking yet another metaphor, casts the achievement into doubt:

Suppose an artist puts forward the fantastic theory that the form of a human head exists in a rough-shaped block of marble. All our rational instinct is roused against such an anthropomorphic speculation. It is inconceivable that nature should have placed such a form inside the block. But the artist proceeds to verify his theory experimentally—with quite rudimentary apparatus too. Merely using a chisel to separate the form for our inspection, he triumphantly proves his theory. (Ibid., 110)

This sort of speculation leads Eddington to wonder, "when the late Lord Rutherford showed us the atomic nucleus, did he *find* it or did he *make* it?" (ibid., 109). Such is the subversive power of Eddington's accumulated metaphors that he can be reasonably sure that his dazzled readers have followed him to his radical conclusion, "Without further apology I shall now assume the reader's assent to the proposition that all the fundamental laws and constants of physics can be deduced unambiguously from a priori considerations and are therefore wholly subjective" (ibid., 62). While this pronouncement clearly makes room for Derrida's "*Il n'y a pas de hors-texte*," Eddington's project was not, and did not pretend to be, entirely deconstructive. Nevertheless, the Newtonian worldview, by now part of the very warp and woof of the culture he addressed, had to be unraveled in order to make way for a new quantum relativistic tapestry. No doubt Eddington's nonscientific audience enjoyed the "humble and stammering" posture of the new philosophy of physics as welcome relief from the "proud and dictatorial" stance of the old. But they may also have relished what A. Lovejoy calls the "metaphysical pathos" that is felt when one has the sense, real or illusory, that one has been allowed to enter the "holy of holies" and to comprehend what has hitherto been a mystery. The problem is that what is mysterious is not always true but the gratitude that is felt upon penetration of a mystery is very similar to that which accompanies the discovery of a truth and may be mistaken for it.

Like Eddington, Derrida and his "postmodern" fellows seek to reconstitute a world which is "always and already" infected with the viruses of racism and sexism. Science, a socially embedded practice, is perhaps the most notable of the achievements of that ailing culture and also the chief source of its power. But it also carries the virus. Little wonder then that scientism has been a chief target of the postmodern assault. On Beer's compelling portrait, the "shy Quaker" Eddington would find much to sympathize with in the postmodern critique and in any case provides much ammunition for its cause.

The central rhetorical problem that Eddington and postmodernists share is this: *How deal with a dominant and defective universe of discourse which must provide, nevertheless, the very foundations of one's own argument.* Eddington's procedure is well illustrated in a passage that Beer quotes at length:

> I am standing on the threshold about to enter a room. It is a complicated business. In the first place I must shove against an atmosphere pressing with a force of fourteen pounds on every square inch of my body. I must make sure of landing on a plank travelling at twenty miles a second round the sun—a fraction of a second too early or too late, the plank would be miles away. I must do this whilst hanging from a round planet head outward into space, and with a wind of aether blowing at no one knows how many miles a second through every interstice of my body. The plank has no solidity of substance. To step on it is like stepping on a swarm of flies. Shall I not slip through? No, if I make the venture one of the flies hits me and gives a boost up again; I may hope that the net result will be that I remain about steady; but if unfortunately I should slip through the floor or be boosted too violently up to the ceiling, the occurrence would be, not a violation of the laws of Nature, but a rare coincidence. These are some of the minor difficulties. (1928, 342)

These would not be "minor difficulties" if the paragraph really described Eddington's frame of mind when he stood on the threshold of a room. It does not! It is not to be supposed that the emminent physicist is ignorant of the Galilean relativity that is a firm guarantee against his missing the flying plank. It would be churlish to object to Eddington's representing as a "rare coincidence" an event that is in fact unprecedented. Nor, granted that one recognizes the obduracy of the commonsense realm he is trying to modify, can one refuse to grant Eddington the pedagogical license to conflate the macro- and the microcosmic in so provocative a fashion. One can ask, however, how can Eddington stand so firmly on the threshold? Is it not the case that the threshold from which he proposes to depart is solidly ensconsed in the Newtonian universe? Is not the only route to the quantum relativistic realm he wishes to display via the Newtonian realm he seeks provisionally to destroy?

Eddington and Virginia Woolf were both born in 1882 well before Rutherford rendered reality porous, or DeBroglie, Schrödinger, and

Born turned it into "Waves, Waves, Waves." Both Eddington and Woolf were in their late teens by the time Planck reluctantly quantized the atom, and both were in their early twenties by the time Einstein relativized space and time. But if their chronologies coincided, their intellectual universes were vastly divergent. He was a senior wrangler in mathematics at Cambridge, she, denied a university education by the sexist society into which she was born, is said to have counted on her fingers all her life. Eddington seems to have fallen heir to the best institutional resources available in math and science. Woolf made brilliant use of her father's library, but that and the occasional tutor seems to be all that her situation allowed.

More poignant was the different psychological temperaments the two contemporaries brought to their oeuvres. By the time the quantum relativistic revolution engulfed science Eddington was thoroughly anchored in the Newtonian universe. He was indeed a shy man, but we may be sure that he never really hesitated on the threshold of a room. The charming and seductive metaphor to the contrary notwithstanding, Eddington possessed not two tables but one, and that single table was rock solid. He dwelt, in short, in the Newtonian world he so brilliantly sought to undermine. And beyond that world, much to Russell's distress, Eddington sensed another spiritual realm at least as real.

Woolf on the contrary appears not to have taken much interest in science until the new physics captured her imagination. Beer recounts the novelist's reaction to a conversation at a supper party, "I wanted, like a child, to stay and argue. True, the argument was passing my limits—how, if Einstein is true, we shall be able to foretell our own lives." As Beer notes this is, of course, a misreading—one that Eddington specifically prohibits. Eddington would perhaps have been more sympathetic with another of Woolf's comments, "It's odd that science, so they tell me, is making things (so to speak) more spiritual." One senses that the ethereal world that Eddington conjured up for pedagogical purposes was, for Woolf, suffering bouts of insanity since her thirteenth year, a psychological reality. Eddington died in 1944, the year before the mighty explosion at Almagordo ushered in the nuclear age. One wonders whether that resounding event, which Eddington and his scientific colleagues had made possible, would have shaken his conviction that the fundamental laws of physics are "wholly subjective." We do know that Woolf, having lost her tenuous grip on reality, had already committed suicide in 1941.

There can be no doubt that the rhetorical critique of science is encouraged by the prevailing disillusionment with science and with the culture that is its chief custodian. Like Eddington we face a universe of discourse that is at once sadly deficient and at the same time thoroughly integrated into our daily lives. The countercultural revolution of the 1960s was a youthful expression of alienation from a technocratic society whose manifest deficiencies are most frequently attributed to science (see Roszak 1969). It is tempting to combat the arrogance of scientistic culture by arguing that if, to paraphrase Orwell, all cultures are equal then it is true that some cultures—notably our own—are less equal. We, like Eddington, can strike such a posture only because we are firmly based in the culture, and well rewarded by it. If we use the omnipresence of "absence," the decided undecideability of things to undermine our culture, we do so well knowing that it is always there to return to. Or, perhaps not.

REFERENCES

Eddington, A. 1928. *The Nature of the Physical World*. New York: MacMillan Company.
———. 1935. *New Pathways in Science*. New York: MacMillan Company.
———. 1939. *The Philosophy of Physical Science*. New York: MacMillan Company.
Roszak, T. 1969. *The Making of a Counter Culture: Reflections on the Technocratic Society and Its Youthful Opposition*. New York. Doubleday & Company.